XSQA

2016 SQA Past Papers With Answers

Higher
PHYSICS

D0673886

2014 Specimen Question Paper,
2015 & 2016 Exams

HODDER GIBSON
AN HACHETTE UK COMPANY

This book contains the official 2014 SQA Specimen Question Paper; 2015 and 2016 Exams for Higher Physics, with associated SQA-approved answers modified from the official marking instructions that accompany the paper.

In addition the book contains study skills advice. This advice has been specially commissioned by Hodder Gibson, and has been written by experienced senior teachers and examiners in line with the new Higher for CfE syllabus and assessment outlines. This is not SQA material but has been devised to provide further practice for Higher examinations.

Hodder Gibson is grateful to the copyright holders, as credited on the final page of the Answer Section, for permission to use their material. Every effort has been made to trace the copyright holders and to obtain their permission for the use of copyright material. Hodder Gibson will be happy to receive information allowing us to rectify any error or omission in future editions.

Hachette UK's policy is to use papers that are natural, renewable and recyclable products and made from wood grown in sustainable forests. The logging and manufacturing processes are expected to conform to the environmental regulations of the country of origin.

Orders: please contact Bookpoint Ltd, 130 Park Drive, Milton Park, Abingdon, Oxon OX14 4SE. Telephone: (44) 01235 827720. Fax: (44) 01235 400454. Lines are open 9.00–5.00, Monday to Saturday, with a 24-hour message answering service. Visit our website at www.hoddereducation.co.uk. Hodder Gibson can be contacted direct on: Tel: 0141 333 4650; Fax: 0141 404 8188; email: hoddergibson@hodder.co.uk

This collection first published in 2016 by
Hodder Gibson, an imprint of Hodder Education,
An Hachette UK Company
211 St Vincent Street
Glasgow G2 5QY

Typeset by Aptara, Inc.

Printed in the UK

A catalogue record for this title is available from the British Library

ISBN: 978-1-4718-9100-7

3 2 1

2017 2016

Introduction

Study Skills – what you need to know to pass exams!

Pause for thought

Many students might skip quickly through a page like this. After all, we all know how to revise. Do you really though?

Think about this:

"IF YOU ALWAYS DO WHAT YOU ALWAYS DO, YOU WILL ALWAYS GET WHAT YOU HAVE ALWAYS GOT."

Do you like the grades you get? Do you want to do better? If you get full marks in your assessment, then that's great! Change nothing! This section is just to help you get that little bit better than you already are.

There are two main parts to the advice on offer here. The first part highlights fairly obvious things but which are also very important. The second part makes suggestions about revision that you might not have thought about but which WILL help you.

Part 1

DOH! It's so obvious but …

Start revising in good time

Don't leave it until the last minute – this will make you panic.

Make a revision timetable that sets out work time AND play time.

Sleep and eat!

Obvious really, and very helpful. Avoid arguments or stressful things too – even games that wind you up. You need to be fit, awake and focused!

Know your place!

Make sure you know exactly **WHEN and WHERE** your exams are.

Know your enemy!

Make sure you know what to expect in the exam.

How is the paper structured?

How much time is there for each question?

What types of question are involved?

Which topics seem to come up time and time again?

Which topics are your strongest and which are your weakest?

Are all topics compulsory or are there choices?

Learn by DOING!

There is no substitute for past papers and practice papers – they are simply essential! Tackling this collection of papers and answers is exactly the right thing to be doing as your exams approach.

Part 2

People learn in different ways. Some like low light, some bright. Some like early morning, some like evening or night. Some prefer warm, some prefer cold. But everyone uses their BRAIN and the brain works when it is active. Passive learning – sitting gazing at notes – is the most INEFFICIENT way to learn anything. Below you will find tips and ideas for making your revision more effective and maybe even more enjoyable. What follows gets your brain active, and active learning works!

Activity 1 – Stop and review

Step 1

When you have done no more than 5 minutes of revision reading STOP!

Step 2

Write a heading in your own words which sums up the topic you have been revising.

Step 3

Write a summary of what you have revised in no more than two sentences. Don't fool yourself by saying, "I know it, but I cannot put it into words". That just means you don't know it well enough. If you cannot write your summary, revise that section again, knowing that you must write a summary at the end of it. Many of you will have notebooks full of blue/black ink writing. Many of the pages will not be especially attractive or memorable so try to liven them up a bit with colour as you are reviewing and rewriting. **This is a great memory aid, and memory is the most important thing.**

Activity 2 – Use technology!

Why should everything be written down? Have you thought about "mental" maps, diagrams, cartoons and colour to help you learn? And rather than write down notes, why not record your revision material?

What about having a text message revision session with friends? Keep in touch with them to find out how and what they are revising and share ideas and questions.

Why not make a video diary where you tell the camera what you are doing, what you think you have learned and what you still have to do? No one has to see or hear it, but the process of having to organise your thoughts in a formal way to explain something is a very important learning practice.

Be sure to make use of electronic files. You could begin to summarise your class notes. Your typing might be slow, but it will get faster and the typed notes will be easier to read than the scribbles in your class notes. Try to add different fonts and colours to make your work stand out. You can easily Google relevant pictures, cartoons and diagrams which you can copy and paste to make your work more attractive and **MEMORABLE**.

Activity 3 – This is it. Do this and you will know lots!

Step 1

In this task you must be very honest with yourself! Find the SQA syllabus for your subject (www.sqa.org.uk). Look at how it is broken down into main topics called MANDATORY knowledge. That means stuff you MUST know.

Step 2

BEFORE you do ANY revision on this topic, write a list of everything that you already know about the subject. It might be quite a long list but you only need to write it once. It shows you all the information that is already in your long-term memory so you know what parts you do not need to revise!

Step 3

Pick a chapter or section from your book or revision notes. Choose a fairly large section or a whole chapter to get the most out of this activity.

With a buddy, use Skype, Facetime, Twitter or any other communication you have, to play the game "If this is the answer, what is the question?". For example, if you are revising Geography and the answer you provide is "meander", your buddy would have to make up a question like "What is the word that describes a feature of a river where it flows slowly and bends often from side to side?".

Make up 10 "answers" based on the content of the chapter or section you are using. Give this to your buddy to solve while you solve theirs.

Step 4

Construct a wordsearch of at least 10 × 10 squares. You can make it as big as you like but keep it realistic. Work together with a group of friends. Many apps allow you to make wordsearch puzzles online. The words and phrases can go in any direction and phrases can be split. Your puzzle must only contain facts linked to the topic you are revising. Your task is to find 10 bits of information to hide in your puzzle, but you must not repeat information that you used in Step 3. DO NOT show where the words are. Fill up empty squares with random letters. Remember to keep a note of where your answers are hidden but do not show your friends. When you have a complete puzzle, exchange it with a friend to solve each other's puzzle.

Step 5

Now make up 10 questions (not "answers" this time) based on the same chapter used in the previous two tasks. Again, you must find NEW information that you have not yet used. Now it's getting hard to find that new information! Again, give your questions to a friend to answer.

Step 6

As you have been doing the puzzles, your brain has been actively searching for new information. Now write a NEW LIST that contains only the new information you have discovered when doing the puzzles. Your new list is the one to look at repeatedly for short bursts over the next few days. Try to remember more and more of it without looking at it. After a few days, you should be able to add words from your second list to your first list as you increase the information in your long-term memory.

FINALLY! Be inspired...

Make a list of different revision ideas and beside each one write **THINGS I HAVE** tried, **THINGS I WILL** try and **THINGS I MIGHT** try. Don't be scared of trying something new.

And remember – "FAIL TO PREPARE AND PREPARE TO FAIL!"

Higher Physics

Assessment

The examination

Section 1

There are 20 multiple choice questions – ensure you answer all of these, even if it means guessing an answer.

Section 2

This section consists of restricted and extended response questions. The mark scored out of **110** here will be scaled to out of **80** and then added to the mark scored in Section 1, giving a total out of **100**.

The majority of the marks will be awarded for applying **knowledge and understanding**. The other marks will be awarded for **applying scientific inquiry, scientific analytical thinking and problem solving skills**.

The assignment

The evidence will be submitted to SQA for external marking. This will be marked out of **20**.

The final mark for Higher Physics will be out of a total of 120.

The examination – general points

Standard 3-marker

Look out for these. The breakdown of the marks will be:

1 mark – selecting equation

1 mark – substitution

1 mark – answer, including unit.

Do not rearrange equations in algebraic form. Select the appropriate equation, substitute the given values, and then rearrange the equation to obtain the required unknown. **This minimises the risk of wrong substitution.**

For example:

Calculate the acceleration of a mass of 5 kg when acted on by a resultant force of 10 N.

Solution 1	Solution 2	Solution 3
$F = ma$ (1)	$F = ma$ (1)	$F = ma$ (1)
$10 = 5a$ (1)	$a = m/F = 5/10$	$10 = 5a$ (1)
$a = 2\ ms^{-2}$ (1)	$= 0.5\ ms^{-2}$	$= 0.5\ ms^{-2}$
3 marks	1 mark for selecting formula.	2 marks (1 mark for selecting formula 1 mark for correct substitution.)

Use of the data sheet and data booklet

Clearly show where you have substituted a value from the data sheet. Do not leave *Ho* in an equation. You must show the value of *Ho* has been correctly substituted.

Rounding – do not round the given values.

E.g. mass of a proton $= 1.673 \times 10^{-27}$ kg
 NOT 1.67×10^{-27} kg.

Although many of the required equations are given, **it is better to know the basic equations to gain time in the examination**.

"Show" questions

Generally **all steps** for these must be given. **Do not assume that substitutions are obvious to the marker.** All equations used must be stated separately and then clearly substituted if required. Many candidates will look at the end product and somehow end up with the required answer. The marker has to ensure that the path to the solution is clear. It is good practice to state why certain equations are used, explaining the Physics behind them.

Make sure you include the unit in the final answer.

Definitions

Know and understand definitions given in the course. Definitions often come from the interpretation of an equation.

For example, 1 Farad is equivalent to a $1CV^{-1}$ $(Q = CV)$

Diagrams, graphs and sketch graphs

When drawing diagrams, use a ruler and use appropriate labels. Angles will be important in certain diagrams. Too many candidates attempt to draw ray diagrams freehand.

When drawing graphs, use a ruler and pencil to draw for axes. Label the axes correctly including units and the origin.

When tackling sketch graphs, care should be taken to be as neat as possible. Ensure axes are drawn in pencil with a ruler. Also ensure you use a ruler to draw a straight line graph.

Significant figures

Do not round off in intermediate calculations, but round off in the final answer to an appropriate number of figures.

Rounding off to three significant figures in the final answer will generally be acceptable.

Prefixes

Ensure you know all the prefixes required and be able to convert them to the correct power of 10.

"Explain"/descriptive questions

These tend to be done poorly. Ensure all points are covered and read over again in order to check there are no mistakes. Try to be clear and to the point, highlighting the relevant Physics.

Do not use up and down arrows in a description – this may help you in shorthand, but these must be translated to words.

Be aware some answers require justification. No attempt at a justification can mean no marks awarded.

Two or more attempts at an answer

The attempt that the candidate does not want to be considered should be scored out. Otherwise zero marks could be awarded.

Do not be tempted to give extra information that might be incorrect – marks could be deducted for each incorrect piece of information. This might include converting incorrectly m to nm in the last line of an answer, when it is not required.

At the end of the exam, if you have time, quickly go over each answer and make sure you have the correct unit inserted.

Skills

Experimental descriptions/planning

You could well be called on to describe an experimental set up.

> Ensure your description is clear enough for another person to repeat it.
>
> Include a clearly labelled diagram.

Suggested improvements to experimental procedure

Look at the percentage uncertainties in the measured quantities and decide which is most significant. Suggest how the size of this uncertainty could be reduced – do not suggest use better apparatus! It might be better to repeat readings, so random uncertainty is reduced or increase distances to reduce the percentage uncertainty in scale reading. There could be a systematic uncertainty that is affecting all readings. It really depends on the experiment. Use your judgement.

Handling data

Relationships

There are two methods to prove a direct or inverse relationship.

Graphical approach

Plot the graphs with the appropriate x and y values and look for a straight line – better plotted in pencil in case of mistakes. **Do not force a line through the origin!** (A vs B for a direct relationship, C vs I / D for an inverse relationship)

Algebraic approach

If it appears that $A \propto B$ then calculate the value of A / B **for all values**.

> If these show that $A/B = k$ then the relationship holds.

If it appears that $C \propto 1/D$ then calculate the value $C.D$ **for all values**.

> If these show that $C.D = k$ then the relationship holds.

Using the equation of a straight line, $y = mx + c$.

Be aware that the gradient of the line can often lead to required values.

> For example, finding the internal resistance and emf of a cell.
>
> $E = IR + I r = V + I r$
>
> $V = -r I + E$ in the form of $y = mx + c$

By plotting the graph of V against I, the value of the gradient will give $-r$ and the intercept will give E.

Ensure you are clear on how to calculate the gradient of a line.

Unfamiliar content

If you come across unfamiliar content such as an equation or measurements from an unfamiliar experiment – don't panic! Just read the instructions. Relationships between the quantities can be found graphically or algebraically.

Uncertainties

In this section you need to understand the following:

> Systematic, scale reading (analogue and digital) and random uncertainties.
>
> Percentage, absolute uncertainties.

Percentage uncertainty in final answer is taken as the largest percentage uncertainty in the components.

E.g. V = (7.5 ± 0.1) V	I = (0.85 ± 0.05) A
= 7.5V ± 1.3%	= 0.85 A ± 5.8%
R = V / I = 7.5 / 0.85 = 8.8 Ω + 5.8% = (8.8 ± 0.1) Ω	

Open-ended questions

There will generally be two open ended questions in the paper worth 3 marks each. Some candidates look upon these as mini essays. Remember that they are only worth 3 marks and it gives the opportunity to demonstrate knowledge and understanding. However, do not spend too long on these. It might be better to revisit them at the end of the exam.

Some students prefer to use bullet points to highlight the main areas of understanding.

Ensure you reread the question and understand exactly what is being asked.

Once you have written your response, read over it again to ensure it makes sense.

Good luck!

Remember that the rewards for passing Higher Physics are well worth it! Your pass will help you get the future you want for yourself. In the exam, be confident in your own ability. If you're not sure how to answer a question, trust your instincts and just give it a go anyway – keep calm and don't panic! GOOD LUCK!

National
Qualifications
SPECIMEN ONLY

SQ37/H/02

**Physics
Section 1—Questions**

Date — Not applicable

Duration — 2 hours and 30 minutes

Instructions for the completion of Section 1 are given on *Page two* of your question and answer booklet SQ37/H/01.

Record your answers on the answer grid on *Page three* of your question and answer booklet.

Reference may be made to the Data Sheet on *Page two* of this booklet and to the Relationships Sheet SQ37/H/11.

Before leaving the examination room you must give your question and answer booklet to the Invigilator; if you do not, you may lose all the marks for this paper.

DATA SHEET

COMMON PHYSICAL QUANTITIES

Quantity	Symbol	Value	Quantity	Symbol	Value
Speed of light in vacuum	c	$3.00 \times 10^{8}\,\text{m s}^{-1}$	Planck's constant	h	$6.63 \times 10^{-34}\,\text{J s}$
Magnitude of the charge on an electron	e	$1.60 \times 10^{-19}\,\text{C}$	Mass of electron	m_e	$9.11 \times 10^{-31}\,\text{kg}$
Universal Constant of Gravitation	G	$6.67 \times 10^{-11}\,\text{m}^3\,\text{kg}^{-1}\,\text{s}^{-2}$	Mass of neutron	m_n	$1.675 \times 10^{-27}\,\text{kg}$
Gravitational acceleration on Earth	g	$9.8\,\text{m s}^{-2}$	Mass of proton	m_p	$1.673 \times 10^{-27}\,\text{kg}$
Hubble's constant	H_0	$2.3 \times 10^{-18}\,\text{s}^{-1}$			

REFRACTIVE INDICES

The refractive indices refer to sodium light of wavelength 589 nm and to substances at a temperature of 273 K.

Substance	Refractive index	Substance	Refractive index
Diamond	2·42	Water	1·33
Crown glass	1·50	Air	1·00

SPECTRAL LINES

Element	Wavelength/nm	Colour	Element	Wavelength/nm	Colour
Hydrogen	656	Red	Cadmium	644	Red
	486	Blue-green		509	Green
	434	Blue-violet		480	Blue
	410	Violet			
	397	Ultraviolet			
	389	Ultraviolet			
Sodium	589	Yellow			

Lasers		
Element	Wavelength/nm	Colour
Carbon dioxide	9550 10590	Infrared
Helium-neon	633	Red

PROPERTIES OF SELECTED MATERIALS

Substance	Density/kg m⁻³	Melting Point/K	Boiling Point/K
Aluminium	2.70×10^3	933	2623
Copper	8.96×10^3	1357	2853
Ice	9.20×10^2	273	. . . .
Sea Water	1.02×10^3	264	377
Water	1.00×10^3	273	373
Air	1·29	. . . .	. . . .
Hydrogen	9.0×10^{-2}	14	20

The gas densities refer to a temperature of 273 K and a pressure of 1.01×10^5 Pa.

SECTION 1 — 20 marks

Attempt ALL questions

1. A trolley has a constant acceleration of $3\,m\,s^{-2}$. This means that

    A   the distance travelled by the trolley increases by 3 metres per second every second

    B   the displacement of the trolley increases by 3 metres per second every second

    C   the speed of the trolley is $3\,m\,s^{-1}$ every second

    D   the velocity of the trolley is $3\,m\,s^{-1}$ every second

    E   the velocity of the trolley increases by $3\,m\,s^{-1}$ every second.

2. Which of the following velocity-time graphs represents the motion of an object that changes direction?

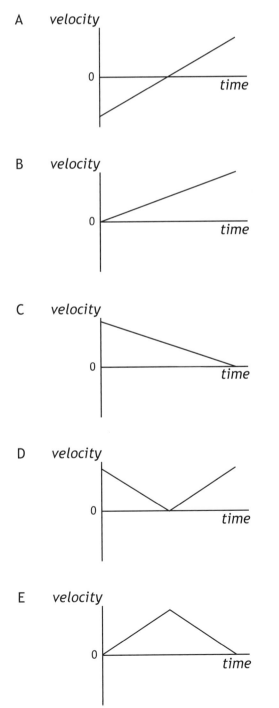

3. A football of mass 0.75 kg is initially at rest. A girl kicks the football and it moves off with an initial speed of 12 m s^{-1}. The time of contact between the girl's foot and the football is 0.15 s.

    The average force applied to the football as it is kicked is

    A 1.4 N

    B 1.8 N

    C 2.4 N

    D 60 N

    E 80 N.

4. Two small asteroids are 12 m apart.

    The masses of the asteroids are 2.0×10^3 kg and 0.050×10^3 kg.

    The gravitational force acting between the asteroids is

    A 1.2×10^{-9} N

    B 4.6×10^{-8} N

    C 5.6×10^{-7} N

    D 1.9×10^{-6} N

    E 6.8×10^3 N.

5. A spaceship on a launch pad is measured to have a length L. This spaceship has a speed of 2.5×10^8 m s^{-1} as it passes a planet.

    Which row in the table describes the length of the spaceship as measured by the pilot in the spaceship and an observer on the planet?

	Length measured by pilot in the spaceship	Length measured by observer on the planet
A	L	less than L
B	L	greater than L
C	L	L
D	less than L	L
E	greater than L	L

6. The siren on an ambulance is emitting sound with a constant frequency of 900 Hz. The ambulance is travelling at a constant speed of 25 m s^{-1} as it approaches and passes a stationary observer. The speed of sound in air is 340 m s^{-1}.

Which row in the table shows the frequency of the sound heard by the observer as the ambulance approaches and as it moves away from the observer?

	Frequency as ambulance approaches (Hz)	Frequency as ambulance moves away (Hz)
A	900	900
B	971	838
C	838	900
D	971	900
E	838	971

7. The photoelectric effect

   A   is evidence for the wave nature of light

   B   can be observed using a diffraction grating

   C   can only be observed with ultra-violet light

   D   can only be observed with infra-red light

   E   is evidence for the particulate nature of light.

8. A ray of red light is incident on a glass block as shown.

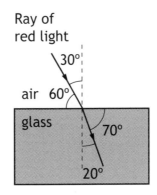

The refractive index of the glass for this light is

   A   0·53

   B   0·68

   C   1·46

   D   1·50

   E   2·53.

9. A ray of red light travels from air into water.

    Which row in the table describes the change, if any, in speed and frequency of a ray of red light as it travels from air into water?

	Speed	Frequency
A	increases	increases
B	increases	stays constant
C	decreases	stays constant
D	decreases	decreases
E	stays constant	decreases

10. Light from a point source is incident on a screen. The screen is $3\cdot0$ m from the source.

    The irradiance at the screen is $8\cdot0$ W m^{-2}.

    The light source is now moved to a distance of 12 m from the screen.

    The irradiance at the screen is now

    A $0\cdot50$ W m^{-2}

    B $1\cdot0$ W m^{-2}

    C $2\cdot0$ W m^{-2}

    D $4\cdot0$ W m^{-2}

    E $8\cdot0$ W m^{-2}.

11. A student makes the following statements about an electron.

    I   An electron is a boson.

    II An electron is a lepton.

    III An electron is a fermion.

    Which of these statements is/are correct?

    A   I only

    B   II only

    C   III only

    D   I and II only

    E   II and III only

12. Radiation of frequency $9 \cdot 40 \times 10^{14}$ Hz is incident on a clean metal surface.

    The work function of the metal is $3 \cdot 78 \times 10^{-19}$ J.

    The maximum kinetic energy of an emitted photoelectron is

    A $2 \cdot 45 \times 10^{-19}$ J

    B $3 \cdot 78 \times 10^{-19}$ J

    C $6 \cdot 23 \times 10^{-19}$ J

    D $1 \cdot 00 \times 10^{-18}$ J

    E $2 \cdot 49 \times 10^{33}$ J.

13. The diagram represents the electric field around a single point charge.

    A student makes the following statements about this diagram.

    I The separation of the field lines indicates the strength of the field.

    II The arrows on the field lines indicate the direction in which an electron would move if placed in the field.

    III The point charge is positive.

    Which of these statements is/are correct?

    A I only

    B II only

    C I and III only

    D II and III only

    E I, II and III

14. In the diagrams below, each resistor has the same resistance.

Which combination has the least value of the effective resistance between the terminals X and Y?

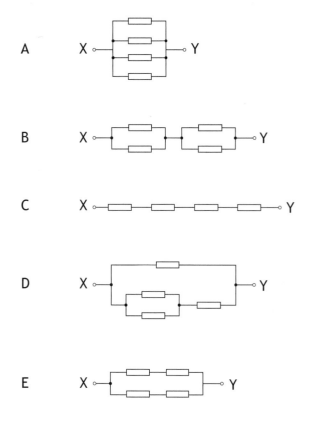

15. A student makes the following statements about charges in electric fields.

   I   An electric field applied to a conductor causes the free electric charges in the conductor to move.

   II When a charge is moved in an electric field work is done.

   III An electric charge experiences a force in an electric field.

Which of these statements is/are correct?

   A   II only
   B   III only
   C   I and II only
   D   II and III only
   E   I, II and III

16. A circuit is set up as shown.

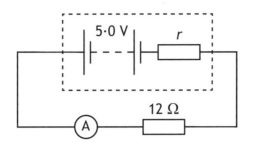

The e.m.f. of the battery is 5·0 V.

The reading on the ammeter is 0·35 A.

The internal resistance r of the battery is

A   0·28 Ω

B   0·80 Ω

C   1·15 Ω

D   2·3 Ω

E   3·2 Ω.

17. The e.m.f. of a battery is

A   the total energy supplied by the battery

B   the voltage lost due to the internal resistance of the battery

C   the total charge that passes through the battery

D   the number of coulombs of charge passing through the battery per second

E   the energy supplied to each coulomb of charge passing through the battery.

18. The r.m.s. voltage of the mains supply is 230 V.

The approximate value of the peak voltage is

A   115 V

B   163 V

C   325 V

D   460 V

E   651 V.

19. Four resistors each of resistance 20 Ω are connected to a 60 V supply of negligible internal resistance as shown.

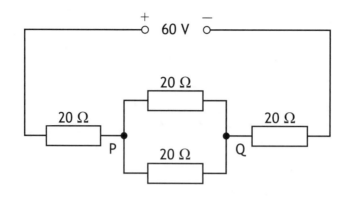

The potential difference across PQ is

A   12 V

B   15 V

C   20 V

D   24 V

E   30 V.

20. Photons with a frequency of $4 \cdot 57 \times 10^{14}$ Hz are incident on a p-n junction in a solar cell. The maximum potential difference these photons produce across this junction is

A   1·34 V

B   1·89 V

C   2·67 V

D   3·79 V

E   5·34 V.

[END OF SECTION 1. NOW ATTEMPT THE QUESTIONS IN SECTION 2 OF YOUR QUESTION AND ANSWER BOOKLET]

National
Qualifications
SPECIMEN ONLY

SQ37/H/11

**Physics
Relationships Sheet**

Date — Not applicable

Relationships required for Physics Higher

$d = \bar{v}t$

$s = \bar{v}t$

$v = u + at$

$s = ut + \frac{1}{2}at^2$

$v^2 = u^2 + 2as$

$s = \frac{1}{2}(u + v)t$

$W = mg$

$F = ma$

$E_W = Fd$

$E_p = mgh$

$E_k = \frac{1}{2}mv^2$

$P = \dfrac{E}{t}$

$p = mv$

$Ft = mv - mu$

$F = G\dfrac{m_1 m_2}{r^2}$

$t' = \dfrac{t}{\sqrt{1 - \left(\frac{v}{c}\right)^2}}$

$l' = l\sqrt{1 - \left(\frac{v}{c}\right)^2}$

$f_o = f_s\left(\dfrac{v}{v \pm v_s}\right)$

$z = \dfrac{\lambda_{observed} - \lambda_{rest}}{\lambda_{rest}}$

$z = \dfrac{v}{c}$

$v = H_0 d$

$W = QV$

$E = mc^2$

$E = hf$

$E_k = hf - hf_0$

$E_2 - E_1 = hf$

$T = \dfrac{1}{f}$

$v = f\lambda$

$d\sin\theta = m\lambda$

$n = \dfrac{\sin\theta_1}{\sin\theta_2}$

$\dfrac{\sin\theta_1}{\sin\theta_2} = \dfrac{\lambda_1}{\lambda_2} = \dfrac{v_1}{v_2}$

$\sin\theta_c = \dfrac{1}{n}$

$I = \dfrac{k}{d^2}$

$I = \dfrac{P}{A}$

path difference $= m\lambda$ or $\left(m + \frac{1}{2}\right)\lambda$ where $m = 0, 1, 2\ldots$

random uncertainty $= \dfrac{\text{max. value} - \text{min. value}}{\text{number of values}}$

$V_{peak} = \sqrt{2}V_{rms}$

$I_{peak} = \sqrt{2}I_{rms}$

$Q = It$

$V = IR$

$P = IV = I^2R = \dfrac{V^2}{R}$

$R_T = R_1 + R_2 + \ldots$

$\dfrac{1}{R_T} = \dfrac{1}{R_1} + \dfrac{1}{R_2} + \ldots$

$E = V + Ir$

$V_1 = \left(\dfrac{R_1}{R_1 + R_2}\right)V_s$

$\dfrac{V_1}{V_2} = \dfrac{R_1}{R_2}$

$C = \dfrac{Q}{V}$

$E = \frac{1}{2}QV = \frac{1}{2}CV^2 = \frac{1}{2}\dfrac{Q^2}{C}$

Additional Relationships

Circle

circumference $= 2\pi r$

area $= \pi r^2$

Sphere

area $= 4\pi r^2$

volume $= \frac{4}{3}\pi r^3$

Trigonometry

$\sin\theta = \dfrac{\text{opposite}}{\text{hypotenuse}}$

$\cos\theta = \dfrac{\text{adjacent}}{\text{hypotenuse}}$

$\tan\theta = \dfrac{\text{opposite}}{\text{adjacent}}$

$\sin^2\theta + \cos^2\theta = 1$

Electron Arrangements of Elements

Key

| Atomic number |
| Symbol |
| Electron arrangement |
| Name |

Group 1 (1)

No.	Symbol	Name	Electron arrangement
1	H	Hydrogen	1
3	Li	Lithium	2,1
11	Na	Sodium	2,8,1
19	K	Potassium	2,8,8,1
37	Rb	Rubidium	2,8,18,8,1
55	Cs	Caesium	2,8,18,18,8,1
87	Fr	Francium	2,8,18,32,18,8,1

Group 2 (2)

No.	Symbol	Name	Electron arrangement
4	Be	Beryllium	2,2
12	Mg	Magnesium	2,8,2
20	Ca	Calcium	2,8,8,2
38	Sr	Strontium	2,8,18,8,2
56	Ba	Barium	2,8,18,18,8,2
88	Ra	Radium	2,8,18,32,18,8,2

Transition Elements

(3)
No.	Symbol	Name	Electron arrangement
21	Sc	Scandium	2,8,9,2
39	Y	Yttrium	2,8,18,9,2
57	La	Lanthanum	2,8,18,18,9,2
89	Ac	Actinium	2,8,18,32,18,9,2

(4)
22	Ti	Titanium	2,8,10,2
40	Zr	Zirconium	2,8,18,10,2
72	Hf	Hafnium	2,8,18,32,10,2
104	Rf	Rutherfordium	2,8,18,32,32,10,2

(5)
23	V	Vanadium	2,8,11,2
41	Nb	Niobium	2,8,18,12,1
73	Ta	Tantalum	2,8,18,32,11,2
105	Db	Dubnium	2,8,18,32,32,11,2

(6)
24	Cr	Chromium	2,8,13,1
42	Mo	Molybdenum	2,8,18,13,1
74	W	Tungsten	2,8,18,32,12,2
106	Sg	Seaborgium	2,8,18,32,32,12,2

(7)
25	Mn	Manganese	2,8,13,2
43	Tc	Technetium	2,8,18,13,2
75	Re	Rhenium	2,8,18,32,13,2
107	Bh	Bohrium	2,8,18,32,32,13,2

(8)
26	Fe	Iron	2,8,14,2
44	Ru	Ruthenium	2,8,18,15,1
76	Os	Osmium	2,8,18,32,14,2
108	Hs	Hassium	2,8,18,32,32,14,2

(9)
27	Co	Cobalt	2,8,15,2
45	Rh	Rhodium	2,8,18,16,1
77	Ir	Iridium	2,8,18,32,15,2
109	Mt	Meitnerium	2,8,18,32,32,15,2

(10)
28	Ni	Nickel	2,8,16,2
46	Pd	Palladium	2,8,18,18,0
78	Pt	Platinum	2,8,18,32,17,1
110	Ds	Darmstadtium	2,8,18,32,32,17,1

(11)
29	Cu	Copper	2,8,18,1
47	Ag	Silver	2,8,18,18,1
79	Au	Gold	2,8,18,32,18,1
111	Rg	Roentgenium	2,8,18,32,32,18,1

(12)
30	Zn	Zinc	2,8,18,2
48	Cd	Cadmium	2,8,18,18,2
80	Hg	Mercury	2,8,18,32,18,2
112	Cn	Copernicium	2,8,18,32,32,18,2

Group 3 (13)
No.	Symbol	Name	Electron arrangement
5	B	Boron	2,3
13	Al	Aluminium	2,8,3
31	Ga	Gallium	2,8,18,3
49	In	Indium	2,8,18,18,3
81	Tl	Thallium	2,8,18,32,18,3

Group 4 (14)
6	C	Carbon	2,4
14	Si	Silicon	2,8,4
32	Ge	Germanium	2,8,18,4
50	Sn	Tin	2,8,18,18,4
82	Pb	Lead	2,8,18,32,18,4

Group 5 (15)
7	N	Nitrogen	2,5
15	P	Phosphorus	2,8,5
33	As	Arsenic	2,8,18,5
51	Sb	Antimony	2,8,18,18,5
83	Bi	Bismuth	2,8,18,32,18,5

Group 6 (16)
8	O	Oxygen	2,6
16	S	Sulfur	2,8,6
34	Se	Selenium	2,8,18,6
52	Te	Tellurium	2,8,18,18,6
84	Po	Polonium	2,8,18,32,18,6

Group 7 (17)
9	F	Fluorine	2,7
17	Cl	Chlorine	2,8,7
35	Br	Bromine	2,8,18,7
53	I	Iodine	2,8,18,18,7
85	At	Astatine	2,8,18,32,18,7

Group 0 (18)
2	He	Helium	2
10	Ne	Neon	2,8
18	Ar	Argon	2,8,8
36	Kr	Krypton	2,8,18,8
54	Xe	Xenon	2,8,18,18,8
86	Rn	Radon	2,8,18,32,18,8

Lanthanides
No.	Symbol	Name	Electron arrangement
57	La	Lanthanum	2,8,18,18,9,2
58	Ce	Cerium	2,8,18,20,8,2
59	Pr	Praseodymium	2,8,18,21,8,2
60	Nd	Neodymium	2,8,18,22,8,2
61	Pm	Promethium	2,8,18,23,8,2
62	Sm	Samarium	2,8,18,24,8,2
63	Eu	Europium	2,8,18,25,8,2
64	Gd	Gadolinium	2,8,18,25,9,2
65	Tb	Terbium	2,8,18,27,8,2
66	Dy	Dysprosium	2,8,18,28,8,2
67	Ho	Holmium	2,8,18,29,8,2
68	Er	Erbium	2,8,18,30,8,2
69	Tm	Thulium	2,8,18,31,8,2
70	Yb	Ytterbium	2,8,18,32,8,2
71	Lu	Lutetium	2,8,18,32,9,2

Actinides
No.	Symbol	Name	Electron arrangement
89	Ac	Actinium	2,8,18,32,18,9,2
90	Th	Thorium	2,8,18,32,18,10,2
91	Pa	Protactinium	2,8,18,32,20,9,2
92	U	Uranium	2,8,18,32,21,9,2
93	Np	Neptunium	2,8,18,32,22,9,2
94	Pu	Plutonium	2,8,18,32,24,8,2
95	Am	Americium	2,8,18,32,25,8,2
96	Cm	Curium	2,8,18,32,25,9,2
97	Bk	Berkelium	2,8,18,32,27,8,2
98	Cf	Californium	2,8,18,32,28,8,2
99	Es	Einsteinium	2,8,18,32,29,8,2
100	Fm	Fermium	2,8,18,32,30,8,2
101	Md	Mendelevium	2,8,18,32,31,8,2
102	No	Nobelium	2,8,18,32,32,8,2
103	Lr	Lawrencium	2,8,18,32,32,9,2

FOR OFFICIAL USE

National
Qualifications
SPECIMEN ONLY

Mark

SQ37/H/01

**Physics Section 1—
Answer Grid and
Section 2**

Date — Not applicable

Duration — 2 hours 30 minutes

Fill in these boxes and read what is printed below.

Full name of centre

Town

Forename(s)

Surname

Number of seat

Date of birth

Day	Month	Year
D D	M M	Y Y

Scottish candidate number

Total marks — 130

SECTION 1 — 20 marks
Attempt ALL questions.
Instructions for the completion of Section 1 are given on *Page two*.

SECTION 2 — 110 marks
Attempt ALL questions.
Reference may be made to the Data Sheet on *Page two* of the question paper SQ37/H/02 and to the Relationship Sheet SQ37/H/11.

Write your answers clearly in the spaces provided in this booklet. Additional space for answers and rough work is provided at the end of this booklet. If you use this space you must clearly identify the question number you are attempting. Any rough work must be written in this booklet. You should score through your rough work when you have written your final copy.

Use **blue** or **black** ink.

Care should be taken to give an appropriate number of significant figures in the final answers to calculations.

Before leaving the examination room you must give this booklet to the Invigilator; if you do not, you may lose all the marks for this paper.

SECTION 1 — 20 marks

The questions for Section 1 are contained in the question paper SQ37/H/02.
Read these and record your answers on the answer grid on *Page three* opposite.
Do **NOT** use gel pens.

1. The answer to each question is **either** A, B, C, D or E. Decide what your answer is, then fill in the appropriate bubble (see sample question below).

2. There is **only one correct** answer to each question.

3. Any rough working should be done on the additional space for answers and rough work at the end of this booklet.

Sample Question

The energy unit measured by the electricity meter in your home is the:

    A   ampere

    B   kilowatt-hour

    C   watt

    D   coulomb

    E   volt.

The correct answer is **B**—kilowatt-hour. The answer **B** bubble has been clearly filled in (see below).

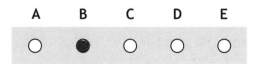

Changing an answer

If you decide to change your answer, cancel your first answer by putting a cross through it (see below) and fill in the answer you want. The answer below has been changed to **D**.

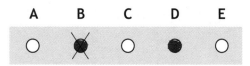

If you then decide to change back to an answer you have already scored out, put a tick (✓) to the **right** of the answer you want, as shown below:

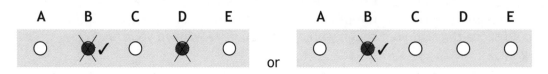

              or

SECTION 1 — Answer Grid

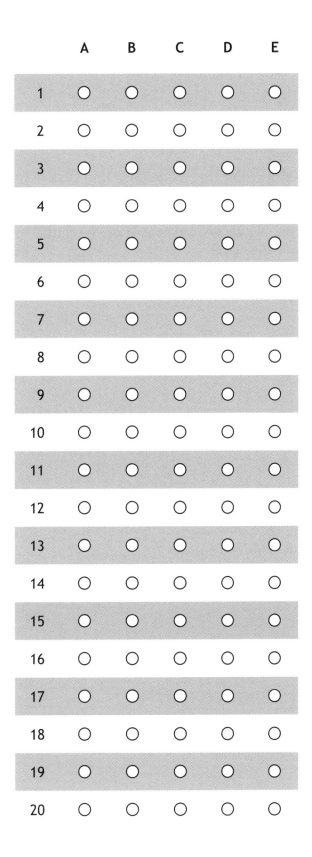

SECTION 2 — 110 marks

Attempt ALL questions

1. A golf ball is hit with a velocity of $50 \cdot 0 \, \text{m s}^{-1}$ at an angle of $35°$ to the horizontal as shown.

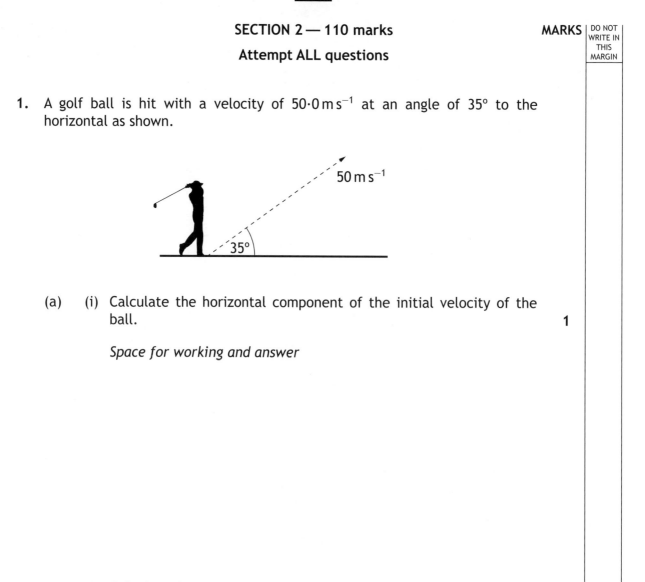

(a) (i) Calculate the horizontal component of the initial velocity of the ball. **1**

Space for working and answer

(ii) Calculate the vertical component of the initial velocity of the ball. **1**

Space for working and answer

MARKS | DO NOT WRITE IN THIS MARGIN

1.   (continued)

(b) The diagram below shows the trajectory of the ball when air resistance is negligible.

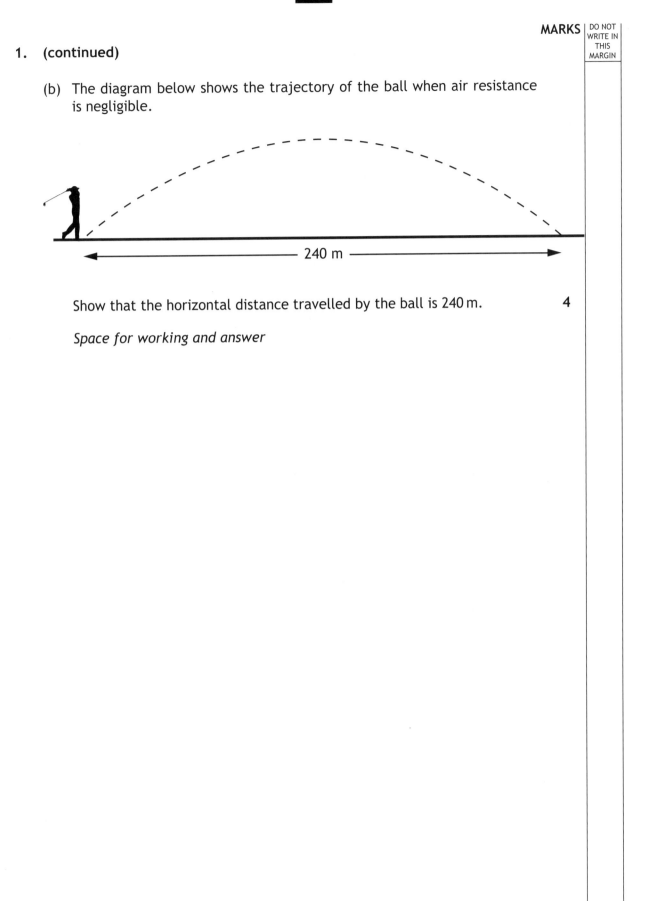

Show that the horizontal distance travelled by the ball is 240 m.   4

Space for working and answer

MARKS | DO NOT WRITE IN THIS MARGIN

2. An electric cart and driver accelerate up a slope. The slope is at an angle of 3·2° to the horizontal. The combined mass of the cart and driver is 220 kg.

(a) (i) Show that the component of the weight of the cart and driver acting down the slope is 120 N.   **2**

Space for working and answer

(ii) At one point on the slope the driving force produced by the cart's motor is 230 N and at this point the total frictional force acting on the cart and driver is 48 N.

Calculate the acceleration of the cart and the driver at this point.   **4**

Space for working and answer

MARKS

2. (a) (continued)

(iii) Explain, in terms of the forces, why there is a maximum angle of slope that the cart can ascend.

2

(b) The electric motor in the cart is connected to a battery of e.m.f. 48 V and internal resistance $0.52\,\Omega$.

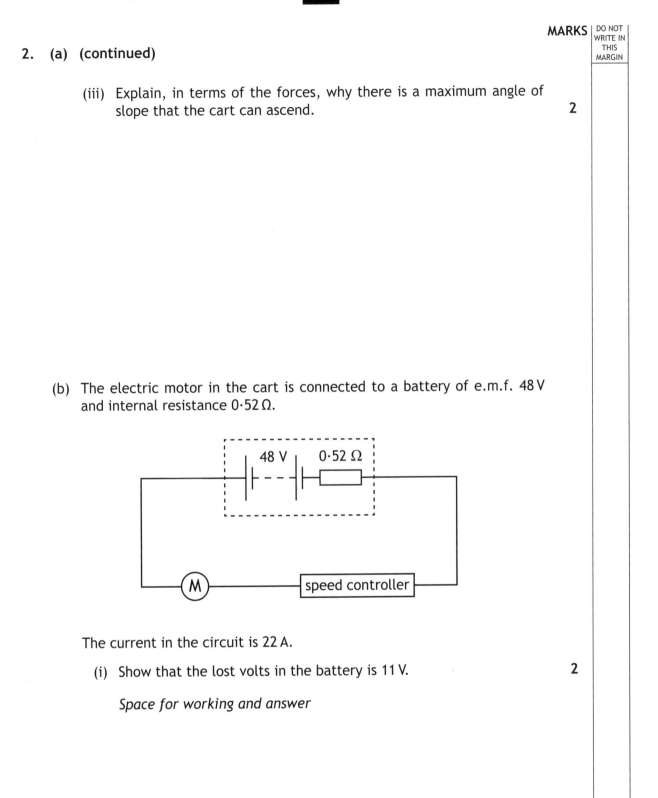

The current in the circuit is 22 A.

(i) Show that the lost volts in the battery is 11 V.

2

Space for working and answer

MARKS | DO NOT WRITE IN THIS MARGIN

2. **(b)** **(continued)**

  (ii) Calculate the output power supplied to the circuit when the current is 22 A.

  4

  Space for working and answer

  (c) The driving force produced by the cart's motor is now increased.

  State what happens to the potential difference across the battery.

  You must justify your answer.

  3

MARKS | DO NOT WRITE IN THIS MARGIN

3. When a car brakes, kinetic energy is turned into heat and sound.

In order to make cars more efficient some manufacturers are developing kinetic energy recovery systems (KERS). These systems store some of the energy that would otherwise be lost as heat and sound.

Estimate the maximum energy that could be stored in such a system when a car brakes.

Clearly show your working for the calculation and any estimates you have made.   **4**

Space for working and answer

MARKS | DO NOT WRITE IN THIS MARGIN

4. Muons are sub-atomic particles produced when cosmic rays enter the atmosphere about 10 km above the surface of the Earth.

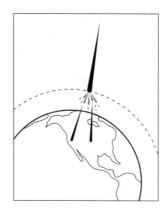

Muons have a mean lifetime of $2 \cdot 2 \times 10^{-6}$ s in their frame of reference. Muons are travelling at $0 \cdot 995c$ relative to an observer on Earth.

(a) Show that the mean distance travelled by the muons in their frame of reference is 660 m.

Space for working and answer

2

(b) Calculate the mean lifetime of the muons as measured by the observer on Earth.

Space for working and answer

3

MARKS | DO NOT WRITE IN THIS MARGIN

4. **(continued)**

(c) Explain why a greater number of muons are detected on the surface of the Earth than would be expected if relativistic effects were not taken into account.

1

MARKS | DO NOT WRITE IN THIS MARGIN

5. A picture of a helmet designed to be worn when riding a bicycle is shown.

The bicycle helmet has a hard outer shell and a soft expanded polystyrene foam liner.

Using your knowledge of physics, comment on the suitability of this design for a bicycle helmet.

3

MARKS | DO NOT WRITE IN THIS MARGIN

6. (a) The diagram below represents part of the emission spectra for the element hydrogen.

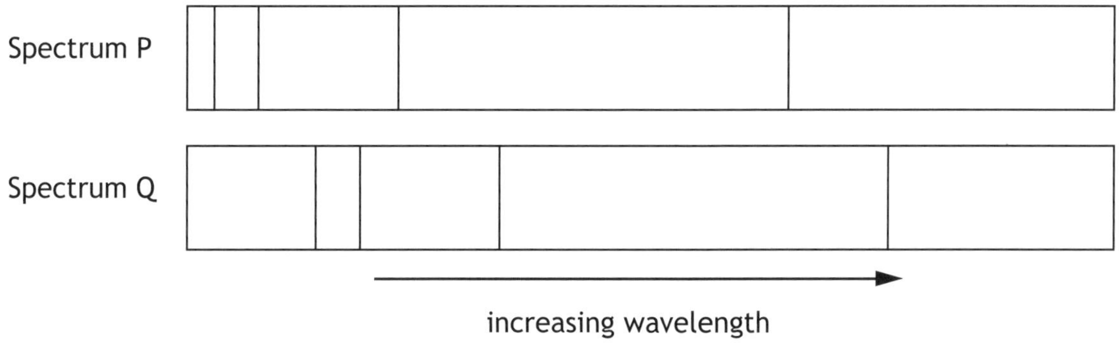

Spectrum P

Spectrum Q

increasing wavelength

Spectrum P is from a laboratory source.

Spectrum Q shows the equivalent lines from a distant star as observed on the Earth.

(i) Explain why spectrum Q is redshifted.   **2**

(ii) One of the lines in spectrum P has a wavelength of 656 nm. The equivalent line in spectrum Q is measured to have a wavelength of 676 nm.

Calculate the recessional velocity of the star.   **5**

Space for working and answer

MARKS | DO NOT WRITE IN THIS MARGIN

6.   (continued)

(b) The recessional velocity of a distant galaxy is $1 \cdot 2 \times 10^7 \, \text{m s}^{-1}$.

Show that the approximate distance to this galaxy is $5 \cdot 2 \times 10^{24} \, \text{m}$.   **2**

Space for working and answer

(c) A student explains the expansion of the Universe using an "expanding balloon model".

The student draws "galaxies" on a balloon and then inflates it.

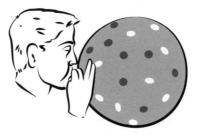

Using your knowledge of physics, comment on the suitability of this model.   **3**

MARKS | DO NOT WRITE IN THIS MARGIN

7. Protons and neutrons are composed of combinations of up and down quarks. Up quarks have a charge of $+\frac{2}{3}e$ while down quarks have a charge of $-\frac{1}{3}e$.

(a) (i) Determine the combination of up and down quarks that makes up:

   (A) a proton;   **1**

   (B) a neutron.   **1**

   (ii) Name the boson that is the mediating particle for the strong force.   **1**

(b) A neutron decays into a proton, an electron and an antineutrino.

$$_{0}^{1}n \rightarrow \, _{1}^{1}p \, + \, _{-1}^{0}e \, + \, \bar{v}$$

Name of this type of decay.   **1**

MARKS | DO NOT WRITE IN THIS MARGIN

8. A linear accelerator is used to accelerate protons.

The accelerator consists of hollow metal tubes placed in a vacuum.

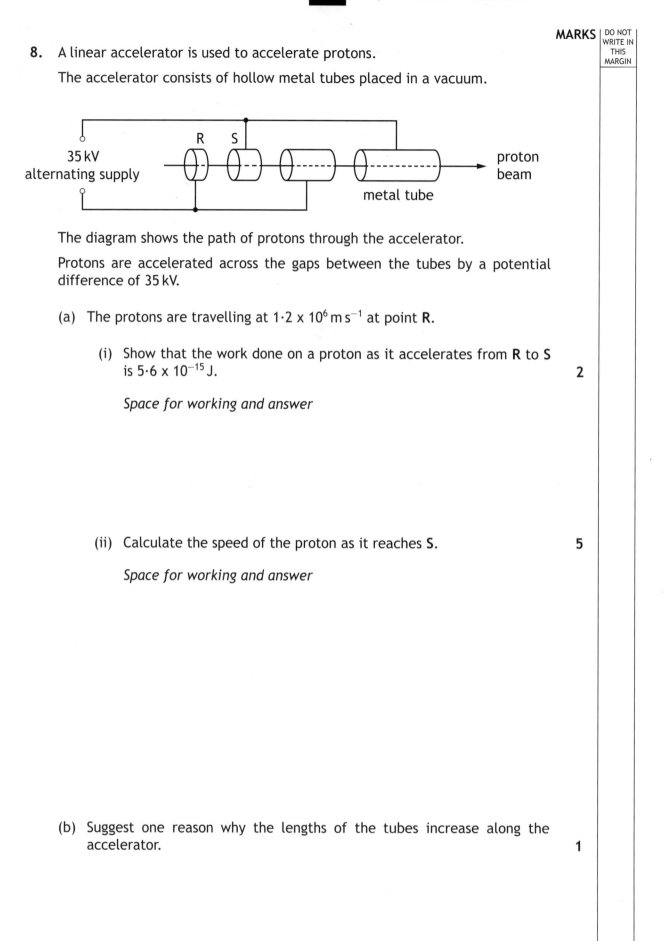

The diagram shows the path of protons through the accelerator.

Protons are accelerated across the gaps between the tubes by a potential difference of 35 kV.

(a) The protons are travelling at $1 \cdot 2 \times 10^6 \, m \, s^{-1}$ at point **R**.

   (i) Show that the work done on a proton as it accelerates from **R** to **S** is $5 \cdot 6 \times 10^{-15} \, J$.   **2**

   Space for working and answer

   (ii) Calculate the speed of the proton as it reaches **S**.   **5**

   Space for working and answer

(b) Suggest one reason why the lengths of the tubes increase along the accelerator.   **1**

MARKS | DO NOT WRITE IN THIS MARGIN

9. (a) The following statement represents a fusion reaction.

$$4\,^1_1\text{H} \rightarrow \,^4_2\text{He} + 2\,^0_1\text{e}^+$$

The masses of the particles involved in the reaction are shown in the table.

Particle	Mass (kg)
^1_1H	$1\cdot673 \times 10^{-27}$
^4_2He	$6\cdot646 \times 10^{-27}$
^0_1e	negligible

(i) Calculate the energy released in this reaction.   **4**

Space for working and answer

(ii) Calculate the energy released when 0·20 kg of hydrogen is converted to helium by this reaction.   **3**

Space for working and answer

MARKS | DO NOT WRITE IN THIS MARGIN

9. (a) (continued)

(iii) Fusion reactors are being developed that use this type of reaction as an energy source.

Explain why this type of fusion reaction is hard to sustain in these reactors.

1

(b) A nucleus of radium-224 decays to radon by emitting an alpha particle.

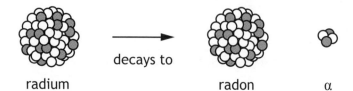

radium   decays to   radon   α

The masses of the particles involved in the decay are shown in the table.

Particle	Mass (kg)
radium-224	$3 \cdot 720 \times 10^{-25}$
radon-220	$3 \cdot 653 \times 10^{-25}$
alpha	$6 \cdot 645 \times 10^{-27}$

Before the decay the radium-224 nucleus is at rest.

After the decay the alpha particle moves off with a velocity of $1 \cdot 460 \times 10^7 \, \text{m s}^{-1}$.

Calculate the velocity of the radon-220 nucleus after the decay.

3

Space for working and answer

MARKS | DO NOT WRITE IN THIS MARGIN

10. The diagram shows equipment used to investigate the photoelectric effect.

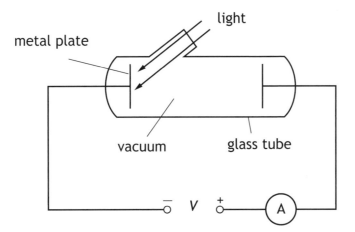

(a) When blue light is shone on the metal plate there is a current in the circuit. When blue light is replaced by red light there is no current.

Explain why this happens.   **2**

(b) The blue light has a frequency of $7 \cdot 0 \times 10^{14}$ Hz.

The work function for the metal plate is $2 \cdot 0 \times 10^{-19}$ J.

Calculate the maximum kinetic energy of the electrons emitted from the plate by this light.   **3**

Space for working and answer

MARKS | DO NOT WRITE IN THIS MARGIN

11. A helium-neon laser produces a beam of coherent red light.

(a) State what is meant by *coherent light*. 1

(b) A student directs this laser beam onto a double slit arrangement as shown in the diagram.

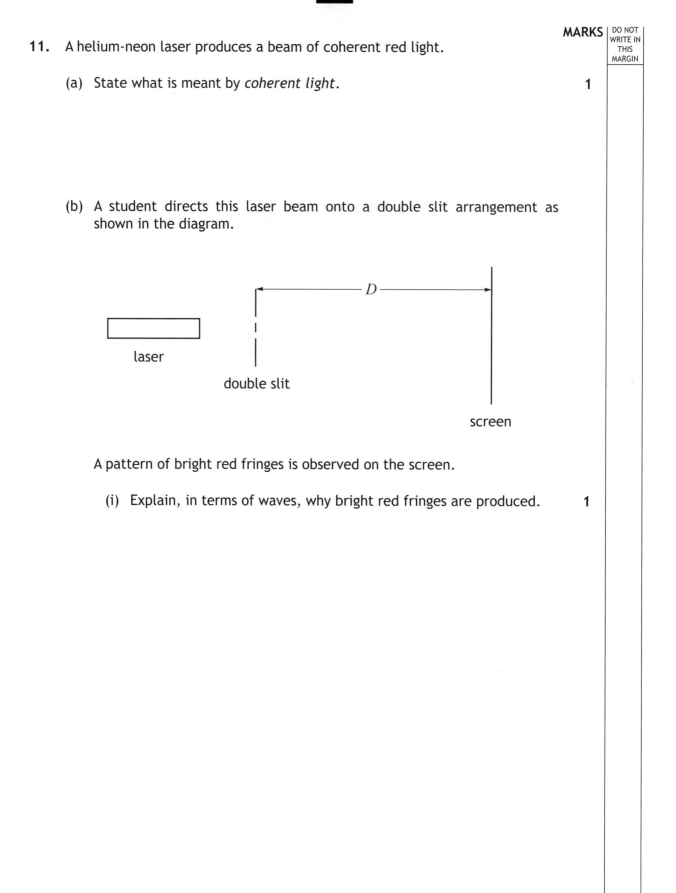

A pattern of bright red fringes is observed on the screen.

(i) Explain, in terms of waves, why bright red fringes are produced. 1

MARKS | DO NOT WRITE IN THIS MARGIN

11. (b) (continued)

(ii) The average separation, Δx, between adjacent fringes is given by the relationship

$$\Delta x = \frac{\lambda D}{d}$$

where: λ is the wavelength of the light
        D is the distance between the double slit and the screen
        d is the distance between the two slits

The diagram shows the value measured by the student of the distance between a series of fringes and the uncertainty in this measurement.

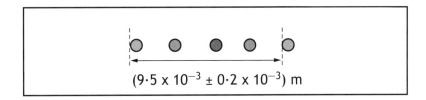

$$(9{\cdot}5 \times 10^{-3} \pm 0{\cdot}2 \times 10^{-3})\ \text{m}$$

The student measures the distance, D, between the double slit and the screen as $(0{\cdot}750 \pm 0{\cdot}001)\ \text{m}$.

Calculate the best estimate of the distance between the two slits.

An uncertainty in the calculated value is not required.

4

Space for working and answer

11.   (b)   (continued)

(iii)   The student wishes to determine more precisely the value of the distance between the two slits *d*.

Show, by calculation, which of the student's measurements should be taken more precisely in order to achieve this.

You must indicate clearly which measurement you have identified.   **3**

Space for working and answer

(iv)   The helium-neon laser is replaced by a laser emitting green light. No other changes are made to the experimental set-up.

Explain the effect this change has on the separation of the fringes observed on the screen.   **2**

MARKS | DO NOT WRITE IN THIS MARGIN

12. A student is investigating the refractive index of a Perspex block for red light.

The student directs a ray of red light towards a semicircular Perspex block as shown.

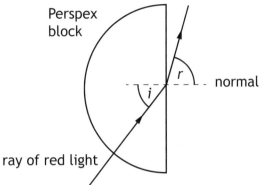

Perspex block

ray of red light

normal

The angle of incidence i is then varied and the angle of refraction r is measured using a protractor.

The following results are obtained.

i (°)	r (°)	sin i	sin r
10	16	0·17	0·28
15	25	0·26	0·42
20	32	0·34	0·53
25	37	0·42	0·60
30	53	0·50	0·80

(a) (i) Using square ruled paper, draw a graph to show how sin r varies with sin i.   3

(ii) Use the graph to determine the refractive index of the Perspex for this light.   2

Space for working and answer

MARKS | DO NOT WRITE IN THIS MARGIN

12. **(a)** **(continued)**

(iii) Suggest **two** ways in which the experimental procedure could be improved to obtain a more accurate value for the refractive index.   **2**

(b) The Perspex block is replaced by an identical glass block with a refractive index of 1·54 and the experiment is repeated.

Determine the maximum angle of incidence that would produce a refracted ray.   **3**

Space for working and answer

MARKS | DO NOT WRITE IN THIS MARGIN

13. A 200 μF capacitor is charged using the circuit shown.

The 12 V battery has negligible internal resistance.

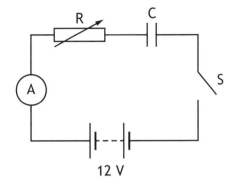

The capacitor is initially uncharged.

The switch S is closed. The charging current is kept constant at 30 μA by adjusting the resistance of the variable resistor, R.

(a) Calculate the resistance of the variable resistor R just after the switch is closed.

Space for working and answer

3

(b) (i) Calculate the charge on the capacitor 30 s after the switch S is closed.

Space for working and answer

3

13. **(b) (continued)**

(ii) Calculate the potential difference across R at this time.

4

Space for working and answer

MARKS | DO NOT WRITE IN THIS MARGIN

14. The electrical conductivity of solids can be explained by band theory.

The diagrams below show the distributions of the valence and conduction bands of materials classified as *conductors*, *insulators* and *semiconductors*.

Shaded areas represent bands occupied by electrons.

The band gap is also indicated.

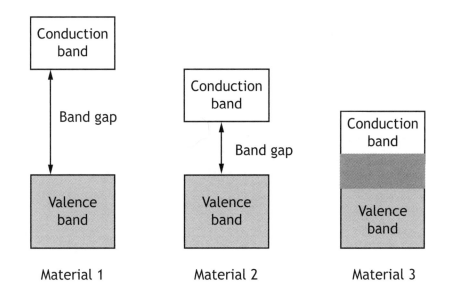

Material 1   Material 2   Material 3

(a) State which material is a semiconductor. 1

(b) A sample of pure semiconductor is heated. Use band theory to explain what happens to the resistance of the sample as it is heated. 2

[END OF SPECIMEN QUESTION PAPER]

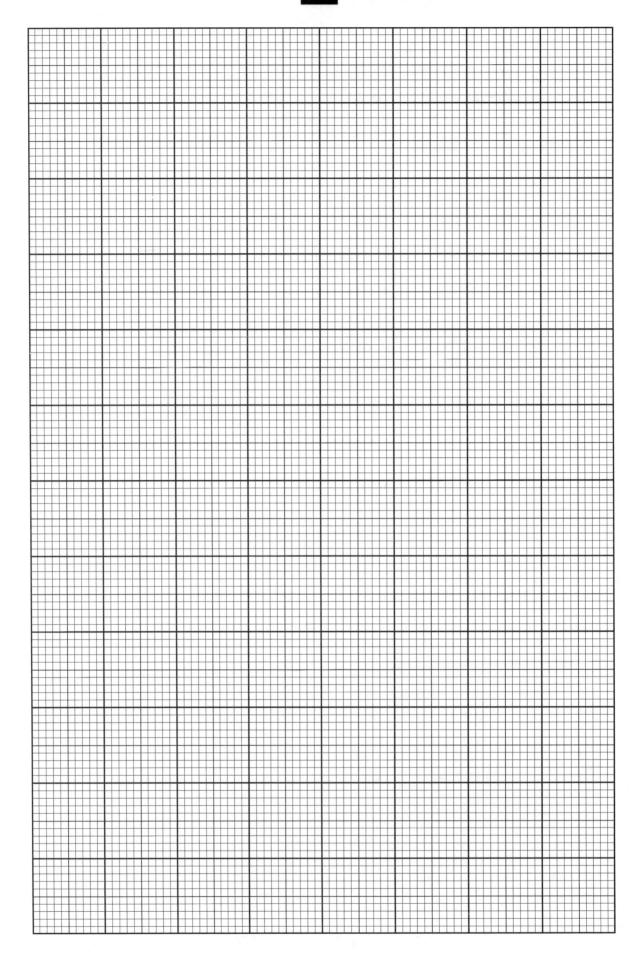

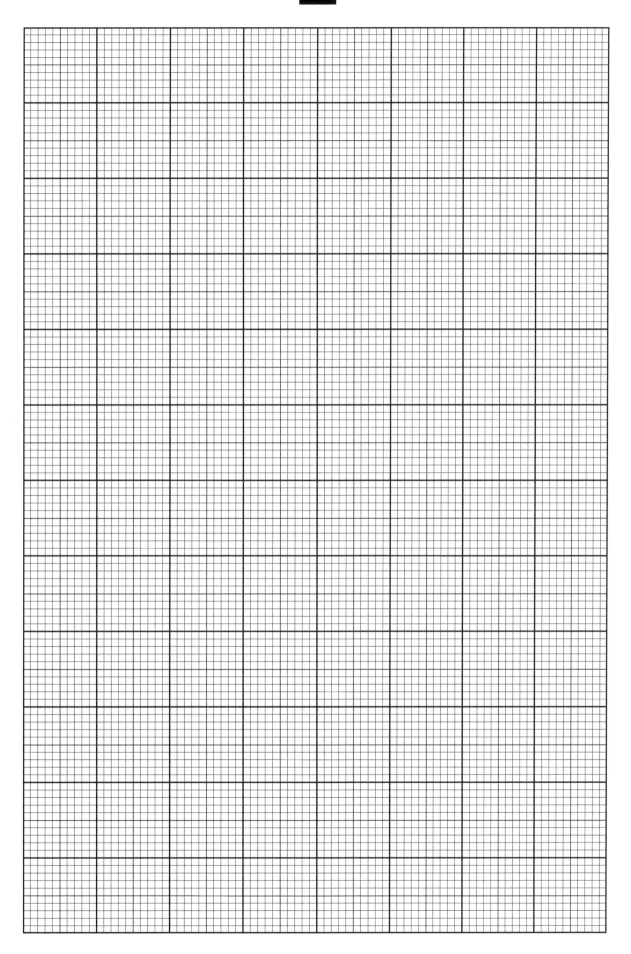

MARKS | DO NOT WRITE IN THIS MARGIN

ADDITIONAL SPACE FOR ANSWERS AND ROUGH WORK

ADDITIONAL SPACE FOR ANSWERS AND ROUGH WORK

[BLANK PAGE]

DO NOT WRITE ON THIS PAGE

HIGHER

2015

National
Qualifications
2015

X757/76/02

Physics
Section 1 — Questions

TUESDAY, 5 MAY
1:00 PM – 3:30 PM

Instructions for the completion of Section 1 are given on *Page two* of your question and answer booklet X757/76/01.

Record your answers on the answer grid on *Page three* of your question and answer booklet.

Reference may be made to the Data Sheet on *Page two* of this booklet and to the Relationships Sheet X757/76/11.

Before leaving the examination room you must give your question and answer booklet to the Invigilator; if you do not, you may lose all the marks for this paper.

DATA SHEET

COMMON PHYSICAL QUANTITIES

Quantity	Symbol	Value	Quantity	Symbol	Value
Speed of light in vacuum	c	$3 \cdot 00 \times 10^{8}\,\mathrm{m\,s^{-1}}$	Planck's constant	h	$6 \cdot 63 \times 10^{-34}\,\mathrm{J\,s}$
Magnitude of the charge on an electron	e	$1 \cdot 60 \times 10^{-19}\,\mathrm{C}$	Mass of electron	m_{e}	$9 \cdot 11 \times 10^{-31}\,\mathrm{kg}$
Universal Constant of Gravitation	G	$6 \cdot 67 \times 10^{-11}\,\mathrm{m^3\,kg^{-1}\,s^{-2}}$	Mass of neutron	m_{n}	$1 \cdot 675 \times 10^{-27}\,\mathrm{kg}$
Gravitational acceleration on Earth	g	$9 \cdot 8\,\mathrm{m\,s^{-2}}$	Mass of proton	m_{p}	$1 \cdot 673 \times 10^{-27}\,\mathrm{kg}$
Hubble's constant	H_{0}	$2 \cdot 3 \times 10^{-18}\,\mathrm{s^{-1}}$			

REFRACTIVE INDICES

The refractive indices refer to sodium light of wavelength 589 nm and to substances at a temperature of 273 K.

Substance	Refractive index	Substance	Refractive index
Diamond	2·42	Water	1·33
Crown glass	1·50	Air	1·00

SPECTRAL LINES

Element	Wavelength/nm	Colour	Element	Wavelength/nm	Colour
Hydrogen	656	Red	Cadmium	644	Red
	486	Blue-green		509	Green
	434	Blue-violet		480	Blue
	410	Violet			

Element	Wavelength/nm	Colour
Carbon dioxide	9550 } 10590 }	Infrared
Helium-neon	633	Red

Hydrogen continued: 397 Ultraviolet, 389 Ultraviolet; Sodium 589 Yellow. Lasers.

PROPERTIES OF SELECTED MATERIALS

Substance	Density/kg m^{-3}	Melting Point/K	Boiling Point/K
Aluminium	$2 \cdot 70 \times 10^{3}$	933	2623
Copper	$8 \cdot 96 \times 10^{3}$	1357	2853
Ice	$9 \cdot 20 \times 10^{2}$	273	. . . .
Sea Water	$1 \cdot 02 \times 10^{3}$	264	377
Water	$1 \cdot 00 \times 10^{3}$	273	373
Air	1·29	. . . .	. . . .
Hydrogen	$9 \cdot 0 \times 10^{-2}$	14	20

The gas densities refer to a temperature of 273 K and a pressure of $1 \cdot 01 \times 10^{5}$ Pa.

SECTION 1 — 20 marks
Attempt ALL questions

1. The following velocity-time graph represents the vertical motion of a ball.

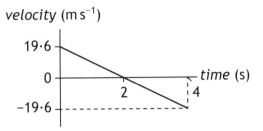

Which of the following acceleration-time graphs represents the same motion?

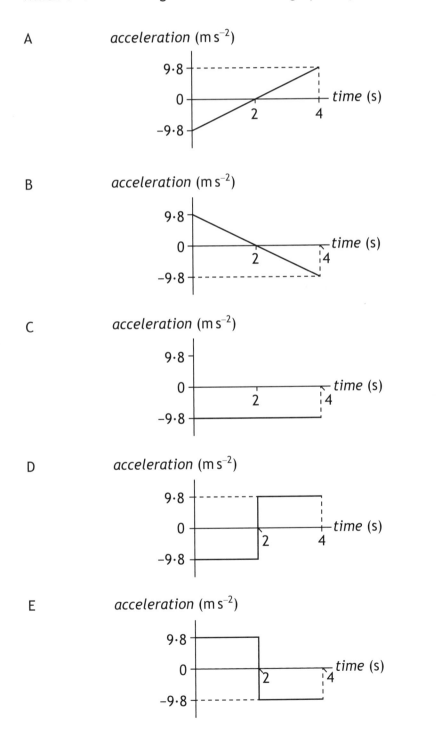

2. A car is travelling at $12\,\text{m}\,\text{s}^{-1}$ along a straight road. The car now accelerates uniformly at $-1\cdot5\,\text{m}\,\text{s}^{-2}$ for $6\cdot0\,\text{s}$.

   The distance travelled during this time is

   A   18 m

   B   45 m

   C   68 m

   D   72 m

   E   99 m.

3. A box of mass m rests on a slope as shown.

   Which row in the table shows the component of the weight acting down the slope and the component of the weight acting normal to the slope?

	Component of weight acting down the slope	Component of weight acting normal to the slope
A	$mg\,\sin\theta$	$mg\,\cos\theta$
B	$mg\,\tan\theta$	$mg\,\sin\theta$
C	$mg\,\cos\theta$	$mg\,\sin\theta$
D	$mg\,\cos\theta$	$mg\,\tan\theta$
E	$mg\,\sin\theta$	$mg\,\tan\theta$

4. A person stands on bathroom scales in a lift.

   The scales show a reading greater than the person's weight.

   The lift is moving

   A   upwards with constant speed

   B   downwards with constant speed

   C   downwards with increasing speed

   D   downwards with decreasing speed

   E   upwards with decreasing speed.

5. A car of mass 900 kg pulls a caravan of mass 400 kg along a straight, horizontal road with an acceleration of $2\cdot0\,m\,s^{-2}$.

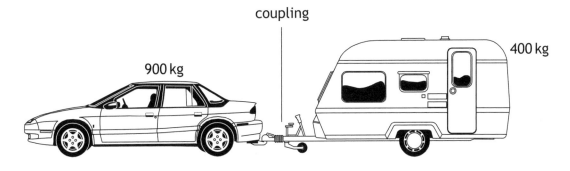

coupling

900 kg

400 kg

Assuming that the frictional forces on the caravan are negligible, the tension in the coupling between the car and the caravan is

A   400 N

B   500 N

C   800 N

D   1800 N

E   2600 N.

6. Water flows at a rate of $6\cdot25 \times 10^8\,kg$ per minute over a waterfall.

The height of the waterfall is 108 m.

The total power delivered by the water in falling through the 108 m is

A $1\cdot13 \times 10^9\,W$

B $1\cdot10 \times 10^{10}\,W$

C $6\cdot62 \times 10^{11}\,W$

D $4\cdot05 \times 10^{12}\,W$

E $3\cdot97 \times 10^{13}\,W$.

7. A spacecraft is travelling at a constant speed of $0\cdot60c$ relative to the Moon.

An observer on the Moon measures the length of the moving spacecraft to be 190 m.

The length of the spacecraft as measured by an astronaut on the spacecraft is

A   120 m

B   152 m

C   238 m

D   297 m

E   300 m.

[Turn over

8. A siren on an ambulance emits sound at a constant frequency of 750 Hz.

The ambulance is travelling at a constant speed of $25 \cdot 0 \, \text{m s}^{-1}$ towards a stationary observer.

The speed of sound in air is $340 \, \text{m s}^{-1}$.

The frequency of the sound heard by the observer is

A   695 Hz

B   699 Hz

C   750 Hz

D   805 Hz

E   810 Hz.

9. The emission of beta particles in radioactive decay is evidence for the existence of

A   quarks

B   electrons

C   gluons

D   neutrinos

E   bosons.

10. Two parallel metal plates X and Y in a vacuum have a potential difference V across them.

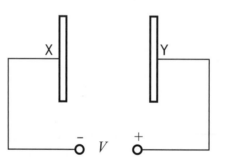

An electron of charge e and mass m, initially at rest, is released from plate X.

The speed of the electron when it reaches plate Y is given by

A $\dfrac{2eV}{m}$

B $\sqrt{\dfrac{2eV}{m}}$

C $\sqrt{\dfrac{2V}{em}}$

D $\dfrac{2V}{em}$

E $\dfrac{2mV}{e}$

11. A potential difference of 2 kV is applied across two metal plates.

An electron passes between the metal plates and follows the path shown.

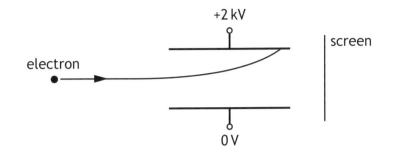

A student makes the following statements about changes that could be made to allow the electron to pass between the plates and reach the screen.

   I   Increasing the initial speed of the electron could allow the electron to reach the screen.

   II Increasing the potential difference across the plates could allow the electron to reach the screen.

   III Reversing the polarity of the plates could allow the electron to reach the screen.

Which of these statements is/are correct?

A   I only

B   II only

C   III only

D   I and II only

E   I and III only

12. The following statement describes a fusion reaction.

$$^{2}_{1}H + {}^{2}_{1}H \circledR {}^{3}_{2}He + {}^{1}_{0}n + \text{energy}$$

The total mass of the particles before the reaction is $6 \cdot 684 \times 10^{-27}$ kg.

The total mass of the particles after the reaction is $6 \cdot 680 \times 10^{-27}$ kg.

The energy released in the reaction is

A $6 \cdot 012 \times 10^{-10}$ J

B $6 \cdot 016 \times 10^{-10}$ J

C $1 \cdot 800 \times 10^{-13}$ J

D $3 \cdot 600 \times 10^{-13}$ J

E $1 \cdot 200 \times 10^{-21}$ J.

[Turn over

13. Two identical loudspeakers, L_1 and L_2, are operated at the same frequency and in phase with each other. An interference pattern is produced.

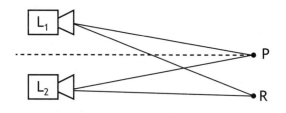

At position P, which is the same distance from both loudspeakers, there is a maximum.

The next maximum is at position R, where $L_1R = 5\cdot6$ m and $L_2R = 5\cdot3$ m.

The speed of sound in air is $340\,\text{m s}^{-1}$.

The frequency of the sound emitted by the loudspeakers is

A $8\cdot8 \times 10^{-4}\,\text{Hz}$

B $3\cdot1 \times 10^{1}\,\text{Hz}$

C $1\cdot0 \times 10^{2}\,\text{Hz}$

D $1\cdot1 \times 10^{3}\,\text{Hz}$

E $3\cdot7 \times 10^{3}\,\text{Hz}$.

14. An experiment is carried out to measure the wavelength of red light from a laser.

The following values for the wavelength are obtained.

650 nm   640 nm   635 nm   648 nm   655 nm

The mean value for the wavelength and the approximate random uncertainty in the mean is

A (645 ± 1) nm

B (645 ± 4) nm

C (646 ± 1) nm

D (646 ± 4) nm

E (3228 ± 20) nm.

15. Red light is used to investigate the critical angle of two materials P and Q.

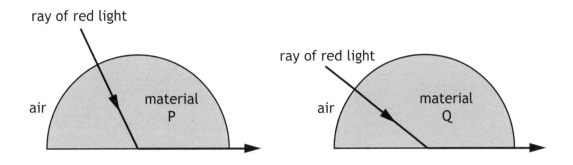

A student makes the following statements.

   I   Material P has a higher refractive index than material Q.

   II The wavelength of the red light is longer inside material P than inside material Q.

   III The red light travels at the same speed inside materials P and Q.

Which of these statements is/are correct?

A   I only

B   II only

C   III only

D   I and II only

E   I, II and III

16. The diagram represents some electron transitions between energy levels in an atom.

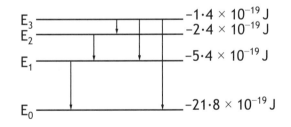

The radiation emitted with the shortest wavelength is produced by an electron making transition

A E_1 to E_0

B E_2 to E_1

C E_3 to E_2

D E_3 to E_1

E E_3 to E_0.

[Turn over

17. The output from a signal generator is connected to the input terminals of an oscilloscope. The trace observed on the oscilloscope screen, the Y-gain setting and the timebase setting are shown.

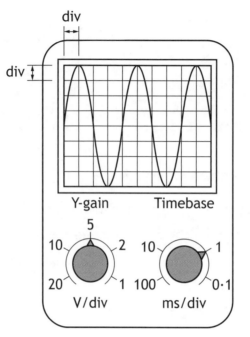

The frequency of the signal shown is calculated using the

A   timebase setting and the vertical height of the trace

B   timebase setting and the horizontal distance between the peaks of the trace

C   Y-gain setting and the vertical height of the trace

D   Y-gain setting and the horizontal distance between the peaks of the trace

E   Y-gain setting and the timebase setting.

18. A circuit is set up as shown.

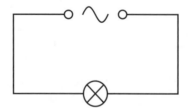

The r.m.s voltage across the lamp is 12 V.

The power produced by the lamp is 24 W.

The peak current in the lamp is

A   0·71 A

B   1·4 A

C   2·0 A

D   2·8 A

E   17 A.

19. A student makes the following statements about energy bands in different materials.

    I   In metals the highest occupied energy band is not completely full.

    II In insulators the highest occupied energy band is full.

    III The gap between the valence band and conduction band is smaller in semiconductors than in insulators.

    Which of these statements is/are correct?

    A   I only

    B   II only

    C   I and II only

    D   I and III only

    E   I, II and III

20. The upward lift force L on the wings of an aircraft is calculated using the relationship

$$L = \tfrac{1}{2}\rho v^2 A C_L$$

    where:

    ρ is the density of air
    v is the speed of the wings through the air
    A is the area of the wings
    C_L is the coefficient of lift.

    The weight of a model aircraft is 80·0 N.
    The area of the wings on the model aircraft is 3·0 m^2.
    The coefficient of lift for these wings is 1·6.
    The density of air is 1·29 kg m^{-3}

    The speed required for the model aircraft to maintain a level flight is

    A   2·5 m s^{-1}

    B   3·6 m s^{-1}

    C   5·1 m s^{-1}

    D   12·9 m s^{-1}

    E   25·8 m s^{-1}.

**[END OF SECTION 1. NOW ATTEMPT THE QUESTIONS IN SECTION 2
OF YOUR QUESTION AND ANSWER BOOKLET]**

[BLANK PAGE]

DO NOT WRITE ON THIS PAGE

**National
Qualifications
2015**

X757/76/11

**Physics
Relationship Sheet**

TUESDAY, 5 MAY
1:00 PM – 3:30 PM

Relationships required for Physics Higher

$d = \bar{v}t$

$s = \bar{v}t$

$v = u + at$

$s = ut + \frac{1}{2}at^2$

$v^2 = u^2 + 2as$

$s = \frac{1}{2}(u + v)t$

$W = mg$

$F = ma$

$E_W = Fd$

$E_p = mgh$

$E_k = \frac{1}{2}mv^2$

$P = \dfrac{E}{t}$

$p = mv$

$Ft = mv - mu$

$F = G\dfrac{m_1 m_2}{r^2}$

$t' = \dfrac{t}{\sqrt{1 - \left(\frac{v}{c}\right)^2}}$

$l' = l\sqrt{1 - \left(\frac{v}{c}\right)^2}$

$f_o = f_s\left(\dfrac{v}{v \pm v_s}\right)$

$z = \dfrac{\lambda_{observed} - \lambda_{rest}}{\lambda_{rest}}$

$z = \dfrac{v}{c}$

$v = H_0 d$

$W = QV$

$E = mc^2$

$E = hf$

$E_k = hf - hf_0$

$E_2 - E_1 = hf$

$T = \dfrac{1}{f}$

$v = f\lambda$

$d\sin\theta = m\lambda$

$n = \dfrac{\sin\theta_1}{\sin\theta_2}$

$\dfrac{\sin\theta_1}{\sin\theta_2} = \dfrac{\lambda_1}{\lambda_2} = \dfrac{v_1}{v_2}$

$\sin\theta_c = \dfrac{1}{n}$

$I = \dfrac{k}{d^2}$

$I = \dfrac{P}{A}$

path difference $= m\lambda$ or $\left(m + \frac{1}{2}\right)\lambda$ where $m = 0, 1, 2 \ldots$

random uncertainty $= \dfrac{\text{max. value} - \text{min. value}}{\text{number of values}}$

$V_{peak} = \sqrt{2}V_{rms}$

$I_{peak} = \sqrt{2}I_{rms}$

$Q = It$

$V = IR$

$P = IV = I^2R = \dfrac{V^2}{R}$

$R_T = R_1 + R_2 + \ldots$

$\dfrac{1}{R_T} = \dfrac{1}{R_1} + \dfrac{1}{R_2} + \ldots$

$E = V + Ir$

$V_1 = \left(\dfrac{R_1}{R_1 + R_2}\right)V_s$

$\dfrac{V_1}{V_2} = \dfrac{R_1}{R_2}$

$C = \dfrac{Q}{V}$

$E = \frac{1}{2}QV = \frac{1}{2}CV^2 = \frac{1}{2}\dfrac{Q^2}{C}$

Additional Relationships

Circle

circumference $= 2\pi r$

area $= \pi r^2$

Sphere

area $= 4\pi r^2$

volume $= \frac{4}{3}\pi r^3$

Trigonometry

$\sin\theta = \dfrac{\text{opposite}}{\text{hypotenuse}}$

$\cos\theta = \dfrac{\text{adjacent}}{\text{hypotenuse}}$

$\tan\theta = \dfrac{\text{opposite}}{\text{adjacent}}$

$\sin^2\theta + \cos^2\theta = 1$

Electron Arrangements of Elements

Key

| Atomic number |
| Symbol |
| Electron arrangement |
| Name |

Transition Elements

| Group 1 | Group 2 | | Group 3 | Group 4 | Group 5 | Group 6 | Group 7 | Group 0 |

Group 1

No.	Symbol	Electron arrangement	Name
1	H	1	Hydrogen
3	Li	2,1	Lithium
11	Na	2,8,1	Sodium
19	K	2,8,8,1	Potassium
37	Rb	2,8,18,8,1	Rubidium
55	Cs	2,8,18,18,8,1	Caesium
87	Fr	2,8,18,32,18,8,1	Francium

Group 2

No.	Symbol	Electron arrangement	Name
4	Be	2,2	Beryllium
12	Mg	2,8,2	Magnesium
20	Ca	2,8,8,2	Calcium
38	Sr	2,8,18,8,2	Strontium
56	Ba	2,8,18,18,8,2	Barium
88	Ra	2,8,18,32,18,8,2	Radium

Transition Elements

Group	No.	Symbol	Electron arrangement	Name
(3)	21	Sc	2,8,9,2	Scandium
(3)	39	Y	2,8,18,9,2	Yttrium
(3)	57	La	2,8,18,18,9,2	Lanthanum
(3)	89	Ac	2,8,18,32,18,9,2	Actinium
(4)	22	Ti	2,8,10,2	Titanium
(4)	40	Zr	2,8,18,10,2	Zirconium
(4)	72	Hf	2,8,18,32,10,2	Hafnium
(4)	104	Rf	2,8,18,32,32,10,2	Rutherfordium
(5)	23	V	2,8,11,2	Vanadium
(5)	41	Nb	2,8,18,12,1	Niobium
(5)	73	Ta	2,8,18,32,11,2	Tantalum
(5)	105	Db	2,8,18,32,32,11,2	Dubnium
(6)	24	Cr	2,8,13,1	Chromium
(6)	42	Mo	2,8,18,13,1	Molybdenum
(6)	74	W	2,8,18,32,12,2	Tungsten
(6)	106	Sg	2,8,18,32,32,12,2	Seaborgium
(7)	25	Mn	2,8,13,2	Manganese
(7)	43	Tc	2,8,18,13,2	Technetium
(7)	75	Re	2,8,18,32,13,2	Rhenium
(7)	107	Bh	2,8,18,32,32,13,2	Bohrium
(8)	26	Fe	2,8,14,2	Iron
(8)	44	Ru	2,8,18,15,1	Ruthenium
(8)	76	Os	2,8,18,32,14,2	Osmium
(8)	108	Hs	2,8,18,32,32,14,2	Hassium
(9)	27	Co	2,8,15,2	Cobalt
(9)	45	Rh	2,8,18,16,1	Rhodium
(9)	77	Ir	2,8,18,32,15,2	Iridium
(9)	109	Mt	2,8,18,32,32,15,2	Meitnerium
(10)	28	Ni	2,8,16,2	Nickel
(10)	46	Pd	2,8,18,18,0	Palladium
(10)	78	Pt	2,8,18,32,17,1	Platinum
(10)	110	Ds	2,8,18,32,32,17,1	Darmstadtium
(11)	29	Cu	2,8,18,1	Copper
(11)	47	Ag	2,8,18,18,1	Silver
(11)	79	Au	2,8,18,32,18,1	Gold
(11)	111	Rg	2,8,18,32,32,18,1	Roentgenium
(12)	30	Zn	2,8,18,2	Zinc
(12)	48	Cd	2,8,18,18,2	Cadmium
(12)	80	Hg	2,8,18,32,18,2	Mercury
(12)	112	Cn	2,8,18,32,32,18,2	Copernicium

Group 3 (13)

No.	Symbol	Electron arrangement	Name
5	B	2,3	Boron
13	Al	2,8,3	Aluminium
31	Ga	2,8,18,3	Gallium
49	In	2,8,18,18,3	Indium
81	Tl	2,8,18,32,18,3	Thallium

Group 4 (14)

No.	Symbol	Electron arrangement	Name
6	C	2,4	Carbon
14	Si	2,8,4	Silicon
32	Ge	2,8,18,4	Germanium
50	Sn	2,8,18,18,4	Tin
82	Pb	2,8,18,32,18,4	Lead

Group 5 (15)

No.	Symbol	Electron arrangement	Name
7	N	2,5	Nitrogen
15	P	2,8,5	Phosphorus
33	As	2,8,18,5	Arsenic
51	Sb	2,8,18,18,5	Antimony
83	Bi	2,8,18,32,18,5	Bismuth

Group 6 (16)

No.	Symbol	Electron arrangement	Name
8	O	2,6	Oxygen
16	S	2,8,6	Sulfur
34	Se	2,8,18,6	Selenium
52	Te	2,8,18,18,6	Tellurium
84	Po	2,8,18,32,18,6	Polonium

Group 7 (17)

No.	Symbol	Electron arrangement	Name
9	F	2,7	Fluorine
17	Cl	2,8,7	Chlorine
35	Br	2,8,18,7	Bromine
53	I	2,8,18,18,7	Iodine
85	At	2,8,18,32,18,7	Astatine

Group 0 (18)

No.	Symbol	Electron arrangement	Name
2	He	2	Helium
10	Ne	2,8	Neon
18	Ar	2,8,8	Argon
36	Kr	2,8,18,8	Krypton
54	Xe	2,8,18,18,8	Xenon
86	Rn	2,8,18,32,18,8	Radon

Lanthanides

No.	Symbol	Electron arrangement	Name
57	La	2,8,18,18,9,2	Lanthanum
58	Ce	2,8,18,20,8,2	Cerium
59	Pr	2,8,18,21,8,2	Praseodymium
60	Nd	2,8,18,22,8,2	Neodymium
61	Pm	2,8,18,23,8,2	Promethium
62	Sm	2,8,18,24,8,2	Samarium
63	Eu	2,8,18,25,8,2	Europium
64	Gd	2,8,18,25,9,2	Gadolinium
65	Tb	2,8,18,27,8,2	Terbium
66	Dy	2,8,18,28,8,2	Dysprosium
67	Ho	2,8,18,29,8,2	Holmium
68	Er	2,8,18,30,8,2	Erbium
69	Tm	2,8,18,31,8,2	Thulium
70	Yb	2,8,18,32,8,2	Ytterbium
71	Lu	2,8,18,32,9,2	Lutetium

Actinides

No.	Symbol	Electron arrangement	Name
89	Ac	2,8,18,32,18,9,2	Actinium
90	Th	2,8,18,32,18,10,2	Thorium
91	Pa	2,8,18,32,20,9,2	Protactinium
92	U	2,8,18,32,21,9,2	Uranium
93	Np	2,8,18,32,22,9,2	Neptunium
94	Pu	2,8,18,32,24,8,2	Plutonium
95	Am	2,8,18,32,25,8,2	Americium
96	Cm	2,8,18,32,25,9,2	Curium
97	Bk	2,8,18,32,27,8,2	Berkelium
98	Cf	2,8,18,32,28,8,2	Californium
99	Es	2,8,18,32,29,8,2	Einsteinium
100	Fm	2,8,18,32,30,8,2	Fermium
101	Md	2,8,18,32,31,8,2	Mendelevium
102	No	2,8,18,32,32,8,2	Nobelium
103	Lr	2,8,18,32,32,9,2	Lawrencium

H

National Qualifications 2015

Mark

X757/76/01

Physics
Section 1 — Answer Grid and Section 2

TUESDAY, 5 MAY

1:00 PM — 3:30 PM

Fill in these boxes and read what is printed below.

Full name of centre

Town

Forename(s)

Surname

Number of seat

Date of birth

Day	Month	Year

Scottish candidate number

Total marks — 130

SECTION 1 — 20 marks
Attempt ALL questions.
Instructions for the completion of Section 1 are given on *Page two*.

SECTION 2 — 110 marks
Attempt ALL questions.

Reference may be made to the Data Sheet on *Page two* of the question paper X757/76/02 and to the Relationship Sheet X757/76/11.

Care should be taken to give an appropriate number of significant figures in the final answers to calculations.

Write your answers clearly in the spaces provided in this booklet. Additional space for answers and rough work is provided at the end of this booklet. If you use this space you must clearly identify the question number you are attempting. Any rough work must be written in this booklet. You should score through your rough work when you have written your final copy.

Use **blue** or **black** ink.

Before leaving the examination room you must give this booklet to the Invigilator; if you do not, you may lose all the marks for this paper.

SECTION 1 — 20 marks

The questions for Section 1 are contained in the question paper X757/76/02.
Read these and record your answers on the answer grid on *Page three* opposite.
Use **blue** or **black** ink. Do NOT use gel pens or pencil.

1. The answer to each question is **either** A, B, C, D or E. Decide what your answer is, then fill in the appropriate bubble (see sample question below).

2. There is **only one correct** answer to each question.

3. Any rough work must be written in the additional space for answers and rough work at the end of this booklet.

Sample Question

The energy unit measured by the electricity meter in your home is the:

    A   ampere

    B   kilowatt-hour

    C   watt

    D   coulomb

    E   volt.

The correct answer is **B**—kilowatt-hour. The answer **B** bubble has been clearly filled in (see below).

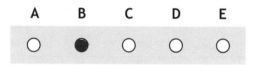

Changing an answer

If you decide to change your answer, cancel your first answer by putting a cross through it (see below) and fill in the answer you want. The answer below has been changed to **D**.

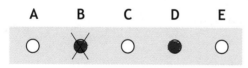

If you then decide to change back to an answer you have already scored out, put a tick (✓) to the **right** of the answer you want, as shown below:

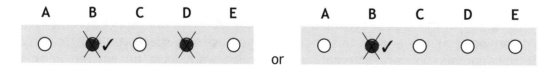

SECTION 1 — Answer Grid

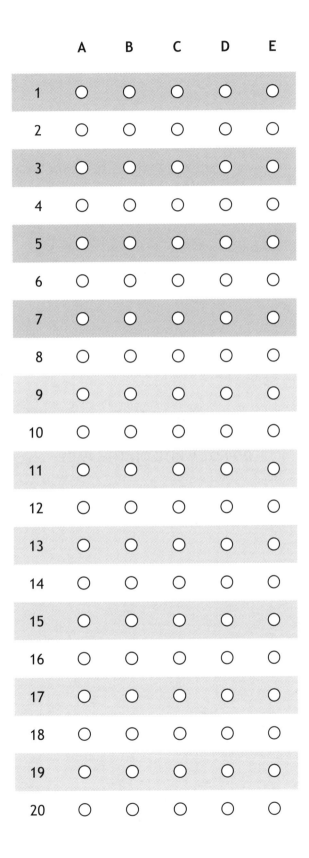

[BLANK PAGE]

DO NOT WRITE ON THIS PAGE

[Turn over for SECTION 2 on *Page six*

DO NOT WRITE ON THIS PAGE

MARKS | DO NOT WRITE IN THIS MARGIN

SECTION 2 — 110 marks

Attempt ALL questions

1. The shot put is an athletics event in which competitors "throw" a shot as far as possible. The shot is a metal ball of mass 4·0 kg. One of the competitors releases the shot at a height of 1·8 m above the ground and at an angle θ to the horizontal. The shot travels through the air and hits the ground at X. The effects of air resistance are negligible.

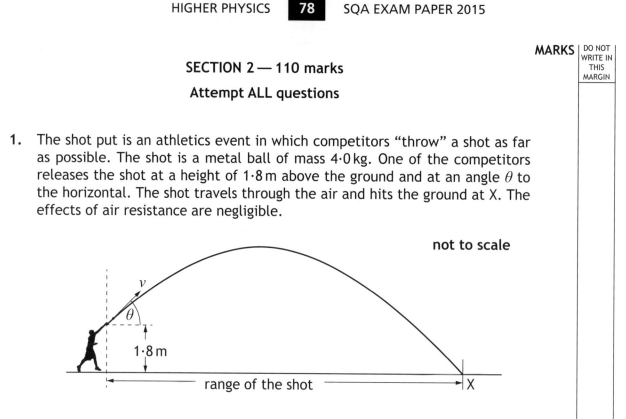

not to scale

The graph shows how the release speed of the shot v varies with the angle of projection θ.

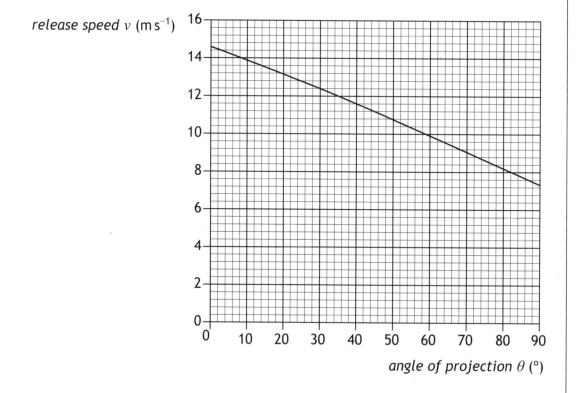

MARKS | DO NOT WRITE IN THIS MARGIN

1. (continued)

(a) The angle of projection for a particular throw is 40°.

   (i) (A) State the release speed of the shot at this angle.   **1**

   (B) Calculate the horizontal component of the initial velocity of the shot.   **1**

   Space for working and answer

   (C) Calculate the vertical component of the initial velocity of the shot.   **1**

   Space for working and answer

   (ii) The maximum height reached by the shot is 4·7 m above the ground. The time between release and reaching this height is 0·76 s.

   (A) Calculate the total time between the shot being released and hitting the ground at X.   **4**

   Space for working and answer

MARKS | DO NOT WRITE IN THIS MARGIN

1. (a) (ii) (continued)

    (B)   Calculate the range of the shot for this throw.   **3**

        Space for working and answer

(b) Using information from the graph, explain the effect of increasing the angle of projection on the kinetic energy of the shot at release.   **2**

1. (a) (ii) (continued)

MARKS

DO NOT WRITE IN THIS MARGIN

2. A student sets up an experiment to investigate collisions between two trolleys on a long, horizontal track.

The mass of trolley X is 0·25 kg and the mass of trolley Y is 0·45 kg.

The effects of friction are negligible.

In one experiment, trolley X is moving at 1·2 m s^{-1} to the right and trolley Y is moving at 0·60 m s^{-1} to the left.

The trolleys collide and do not stick together. After the collision, trolley X rebounds with a velocity of 0·80 m s^{-1} to the left.

(a) Determine the velocity of trolley Y after the collision.

Space for working and answer

3

[Turn over

MARKS | DO NOT WRITE IN THIS MARGIN

2. (continued)

(b) The force sensor measures the force acting on trolley Y during the collision.

The laptop displays the following force-time graph for the collision.

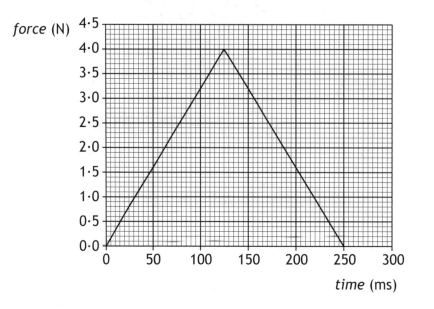

(i) Determine the magnitude of the impulse on trolley Y.   3

Space for working and answer

(ii) Determine the magnitude of the change in momentum of trolley X.   1

MARKS | DO NOT WRITE IN THIS MARGIN

2. (b) (continued)

(iii) Sketch a velocity-time graph to show how the velocity of trolley X varies from 0·50 s before the collision to 0·50 s after the collision.   3

Numerical values are required on both axes.
You may wish to use the square-ruled paper on *Page thirty-six*.

[Turn over

MARKS | DO NOT WRITE IN THIS MARGIN

3. A space probe of mass $5 \cdot 60 \times 10^3$ kg is in orbit at a height of $3 \cdot 70 \times 10^6$ m above the surface of Mars.

space probe

not to scale

Mars

The mass of Mars is $6 \cdot 42 \times 10^{23}$ kg.
The radius of Mars is $3 \cdot 39 \times 10^6$ m.

(a) Calculate the gravitational force between the probe and Mars.　　3

Space for working and answer

(b) Calculate the gravitational field strength of Mars at this height.　　3

Space for working and answer

[Turn over for Question 4 on *Page fourteen*

DO NOT WRITE ON THIS PAGE

MARKS | DO NOT WRITE IN THIS MARGIN

4. Light from the Sun is used to produce a visible spectrum.

A student views this spectrum and observes a number of dark lines as shown.

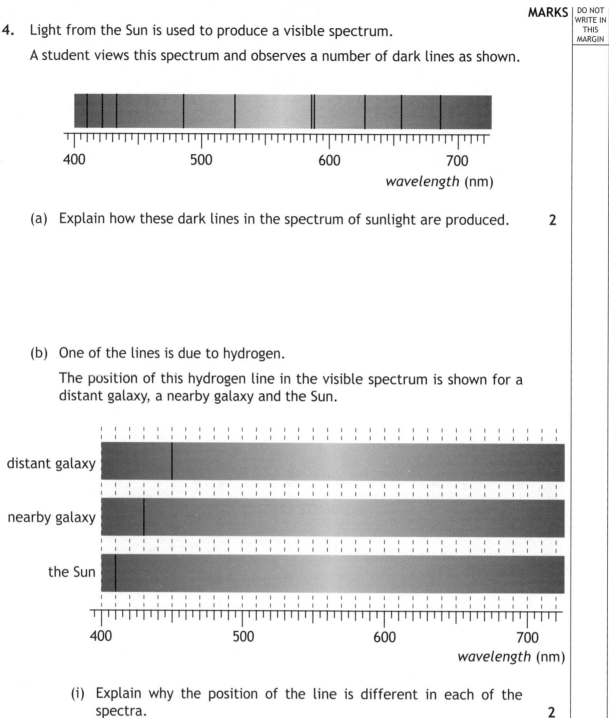

(a) Explain how these dark lines in the spectrum of sunlight are produced.   **2**

(b) One of the lines is due to hydrogen.

The position of this hydrogen line in the visible spectrum is shown for a distant galaxy, a nearby galaxy and the Sun.

(i) Explain why the position of the line is different in each of the spectra.   **2**

MARKS | DO NOT WRITE IN THIS MARGIN

4. **(b)** **(continued)**

    (ii) Show that the redshift of the light from the distant galaxy is 0·098.   **2**

        Space for working and answer

    (iii) Calculate the approximate distance to the distant galaxy.   **5**

        Space for working and answer

[Turn over

5. A quote from a well-known science fiction writer states:

"In the beginning there was nothing, which exploded."

Using your knowledge of physics, comment on the above statement.

3

MARKS

6.　(a)　The Standard Model classifies *force mediating particles* as bosons. Name the boson associated with the electromagnetic force.

1

(b)　In July 2012 scientists at CERN announced that they had found a particle that behaved in the way that they expected the Higgs boson to behave. Within a year this particle was confirmed to be a Higgs boson.

This Higgs boson had a mass-energy equivalence of 126 GeV.
(1 eV = $1 \cdot 6 \times 10^{-19}$ J)

　(i)　Show that the mass of the Higgs boson is $2 \cdot 2 \times 10^{-25}$ kg.

3

Space for working and answer

　(ii)　Compare the mass of the Higgs boson with the mass of a proton in terms of orders of magnitude.

2

Space for working and answer

[Turn over

7. The use of analogies from everyday life can help better understanding of physics concepts. Throwing different balls at a coconut shy to dislodge a coconut is an analogy which can help understanding of the photoelectric effect.

Use your knowledge of physics to comment on this analogy.   3

[Turn over for Question 8 on *Page twenty*

DO NOT WRITE ON THIS PAGE

MARKS | DO NOT WRITE IN THIS MARGIN

8. A student investigates how irradiance I varies with distance d from a point source of light.

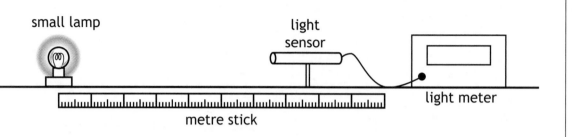

small lamp

light sensor

light meter

metre stick

The distance between a small lamp and a light sensor is measured with a metre stick. The irradiance is measured with a light meter.

The apparatus is set up as shown in a darkened laboratory.

The following results are obtained.

d (m)	0·20	0·30	0·40	0·50
I (W m^{-2})	134·0	60·5	33·6	21·8

(a) State what is meant by the term *irradiance*.   1

(b) Use **all** the data to establish the relationship between irradiance I and distance d .   3

MARKS

8. **(continued)**

(c) The lamp is now moved to a distance of 0·60 m from the light sensor.

Calculate the irradiance of light from the lamp at this distance.　　3

Space for working and answer

(d) Suggest one way in which the experiment could be improved.

You **must** justify your answer.　　2

(e) The student now replaces the lamp with a different small lamp.
The power output of this lamp is 24 W.

Calculate the irradiance of light from this lamp at a distance of 2·0 m.　　4

Space for working and answer

MARKS | DO NOT WRITE IN THIS MARGIN

9. A student carries out two experiments to investigate the spectra produced from a ray of white light.

(a) In the first experiment, a ray of white light is incident on a glass prism as shown.

not to scale

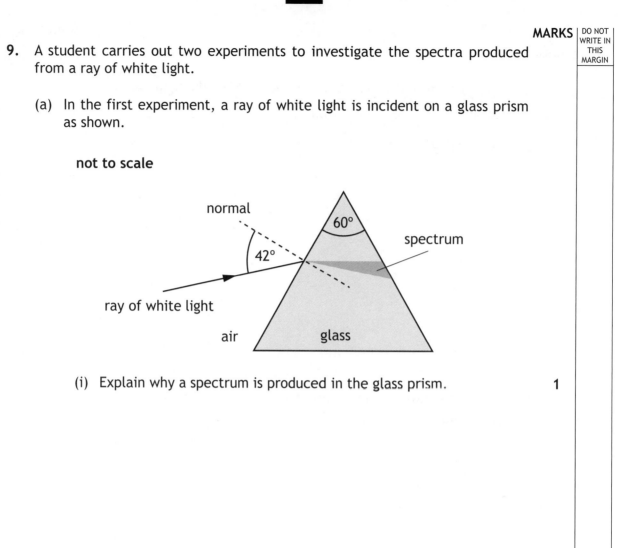

normal

60°

42°

spectrum

ray of white light

air   glass

(i) Explain why a spectrum is produced in the glass prism.   1

(ii) The refractive index of the glass for red light is 1·54.

Calculate the speed of red light in the glass prism.   3

Space for working and answer

MARKS | DO NOT WRITE IN THIS MARGIN

9. **(continued)**

(b) In the second experiment, a ray of white light is incident on a grating.

not to scale

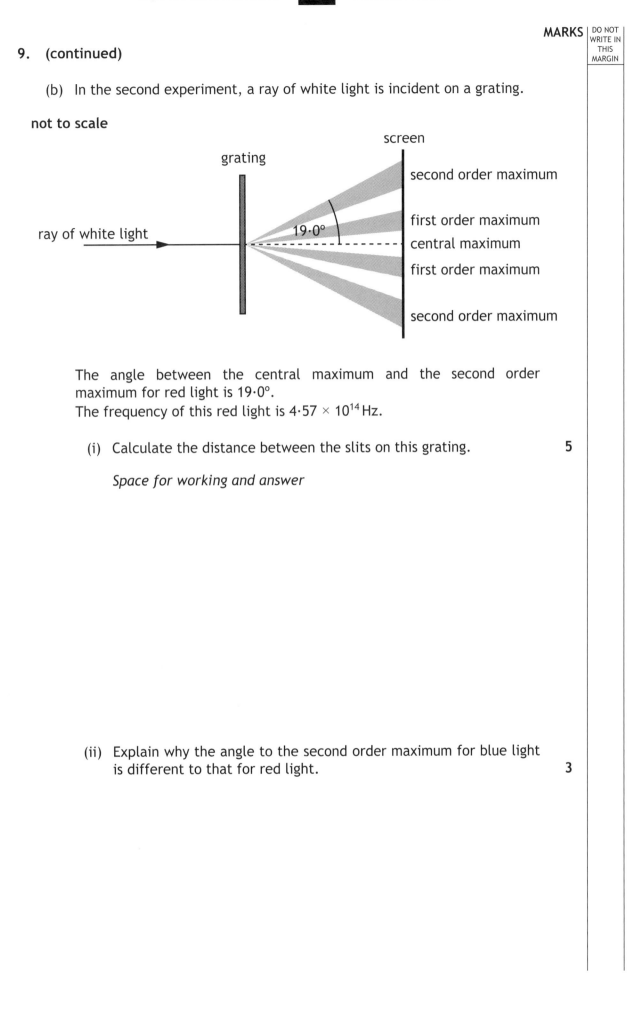

The angle between the central maximum and the second order maximum for red light is 19·0°.
The frequency of this red light is $4·57 \times 10^{14}$ Hz.

(i) Calculate the distance between the slits on this grating.   **5**

Space for working and answer

(ii) Explain why the angle to the second order maximum for blue light is different to that for red light.   **3**

[BLANK PAGE]

DO NOT WRITE ON THIS PAGE

MARKS | DO NOT WRITE IN THIS MARGIN

10. A car battery is connected to an electric motor as shown.

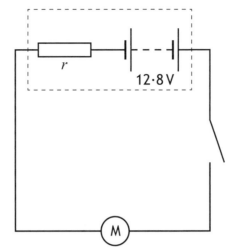

The electric motor requires a large current to operate.

(a) The car battery has an e.m.f. of 12·8 V and an internal resistance r of $6·0 \times 10^{-3}\,\Omega$. The motor has a resistance of 0·050 Ω.

    (i) State what is meant by an *e.m.f. of 12·8 V.*   **1**

    (ii) Calculate the current in the circuit when the motor is operating.   **3**

       Space for working and answer

    (iii) Suggest why the connecting wires used in this circuit have a large diameter.   **1**

MARKS

10. **(continued)**

(b) A technician sets up the following circuit with a different car battery connected to a variable resistor R.

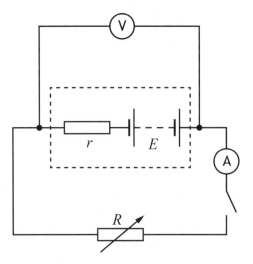

Readings of current I and terminal potential difference V from this circuit are used to produce the following graph.

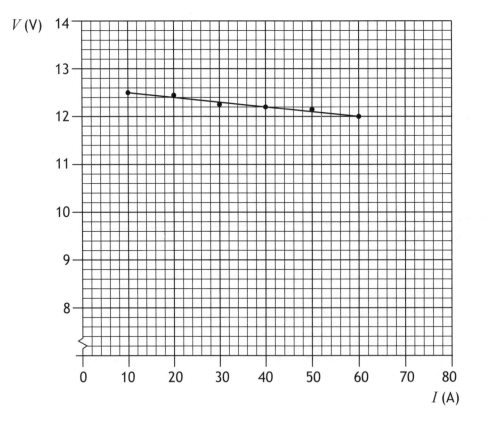

MARKS | DO NOT WRITE IN THIS MARGIN

10. (b) (continued)

Use information from the graph to determine:

(i)   the e.m.f. of the battery;   1

Space for working and answer

(ii) the internal resistance of the battery;   3

Space for working and answer

[Turn over

MARKS | DO NOT WRITE IN THIS MARGIN

10. **(b) (continued)**

(iii) After being used for some time the e.m.f. of the battery decreases to 11·5 V and the internal resistance increases to 0·090 Ω.

The battery is connected to a battery charger of constant e.m.f. 15·0 V and internal resistance of 0·45 Ω as shown.

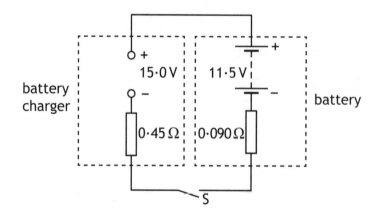

(A) Switch S is closed.

Calculate the initial charging current.   3

Space for working and answer

(B) Explain why the charging current decreases as the battery charges.   2

[Turn over for Question 11 on *Page thirty*]

DO NOT WRITE ON THIS PAGE

MARKS | DO NOT WRITE IN THIS MARGIN

11. A defibrillator is a device that provides a high energy electrical impulse to correct abnormal heart beats.

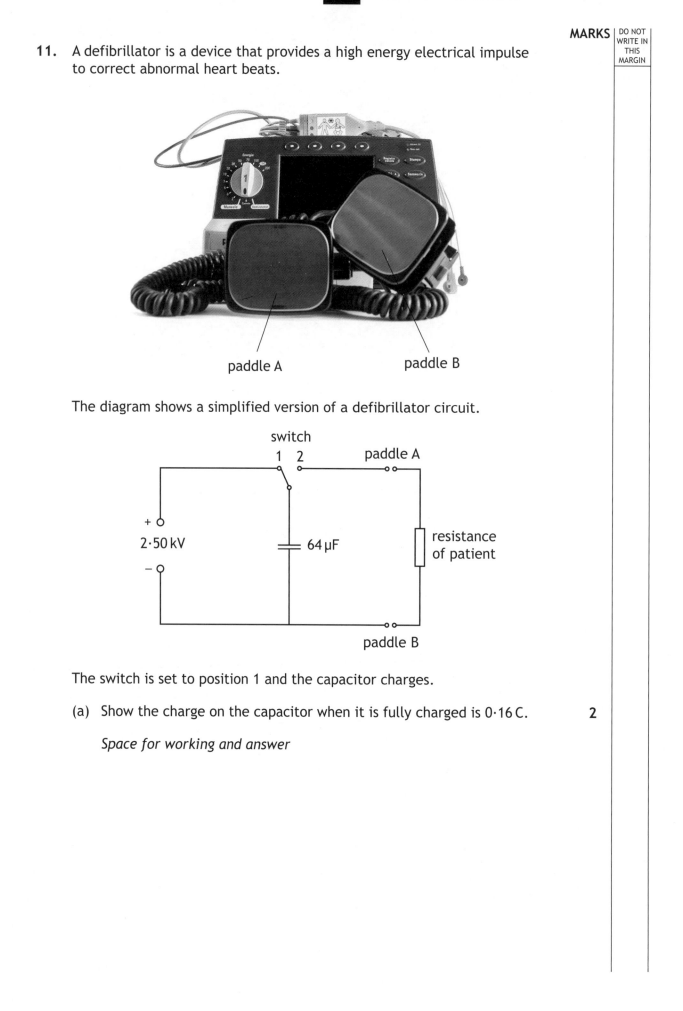

paddle A   paddle B

The diagram shows a simplified version of a defibrillator circuit.

The switch is set to position 1 and the capacitor charges.

(a) Show the charge on the capacitor when it is fully charged is 0·16 C.

Space for working and answer

2

MARKS | DO NOT WRITE IN THIS MARGIN

11. (continued)

(b) Calculate the maximum energy stored by the capacitor.   3

Space for working and answer

(c) To provide the electrical impulse required the capacitor is discharged through the person's chest using the paddles as shown

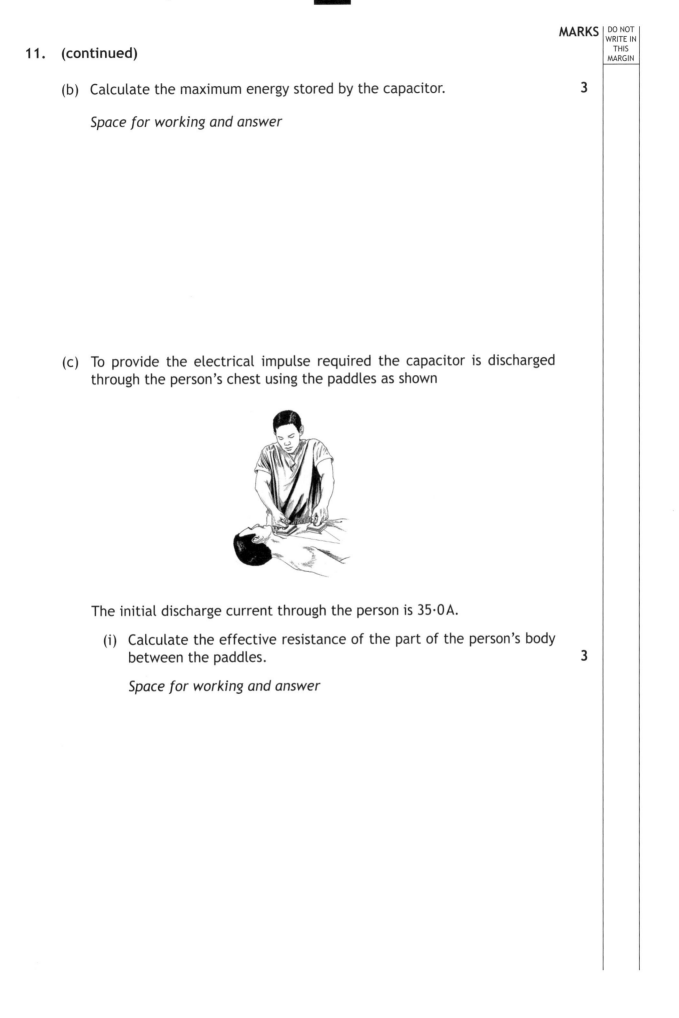

The initial discharge current through the person is 35·0 A.

(i) Calculate the effective resistance of the part of the person's body between the paddles.   3

Space for working and answer

MARKS | DO NOT WRITE IN THIS MARGIN

11. (c) (continued)

(ii) The graph shows how the current between the paddles varies with time during the discharge of the capacitor.

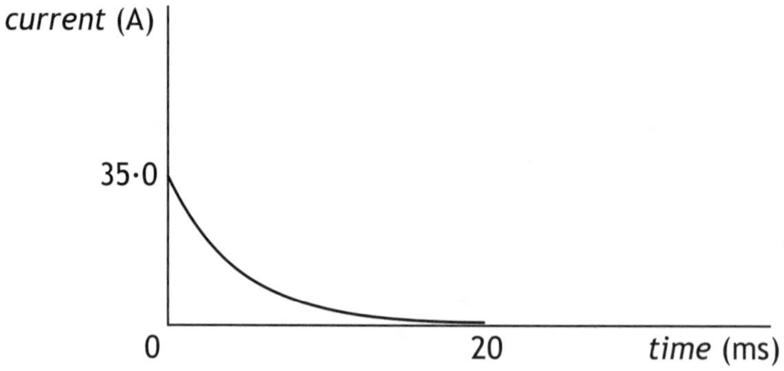

current (A)

35·0

0   20   time (ms)

The effective resistance of the person remains the same during this time.

Explain why the current decreases with time.   1

(iii) The defibrillator is used on a different person with larger effective resistance. The capacitor is again charged to 2·50 kV.

On the graph in (c)(ii) add a line to show how the current in this person varies with time.

(An additional graph, if required, can be found on *Page thirty-eight*).   2

MARKS | DO NOT WRITE IN THIS MARGIN

12. A student carries out an investigation to determine the refractive index of a prism.

A ray of monochromatic light passes through the prism as shown.

not to scale

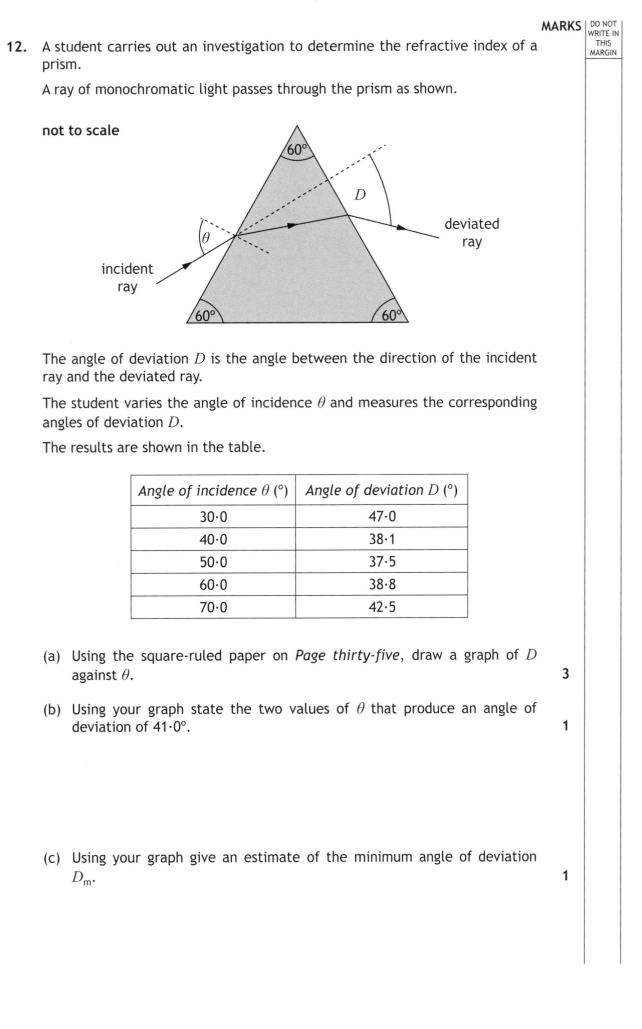

The angle of deviation D is the angle between the direction of the incident ray and the deviated ray.

The student varies the angle of incidence θ and measures the corresponding angles of deviation D.

The results are shown in the table.

Angle of incidence θ (°)	Angle of deviation D (°)
30·0	47·0
40·0	38·1
50·0	37·5
60·0	38·8
70·0	42·5

(a) Using the square-ruled paper on *Page thirty-five*, draw a graph of D against θ.   3

(b) Using your graph state the two values of θ that produce an angle of deviation of 41·0°.   1

(c) Using your graph give an estimate of the minimum angle of deviation D_m.   1

MARKS | DO NOT WRITE IN THIS MARGIN

12. (continued)

(d) The refractive index n of the prism can be determined using the relationship.

$$n \sin\left(\frac{A}{2}\right) = \sin\left(\frac{A + D_m}{2}\right)$$

where A is the angle at the top of the prism, and
D_m is the minimum angle of deviation.

Use this relationship and your answer to (c) to determine the refractive index of the prism.   **2**

Space for working and answer

(e) Using the same apparatus, the student now wishes to determine more precisely the minimum angle of deviation.

Suggest two improvements to the experimental procedure that would achieve this.   **2**

[END OF QUESTION PAPER]

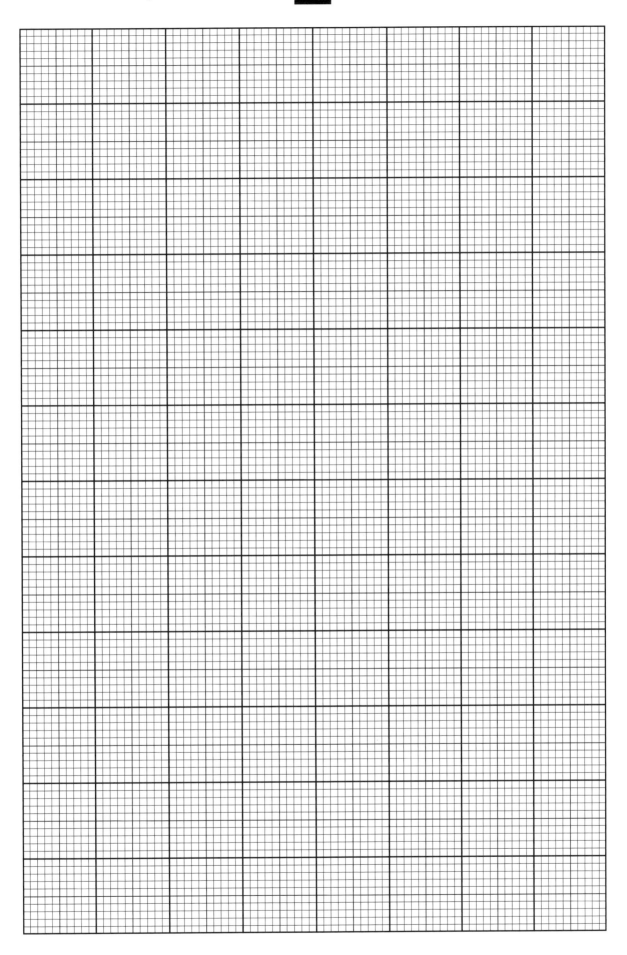

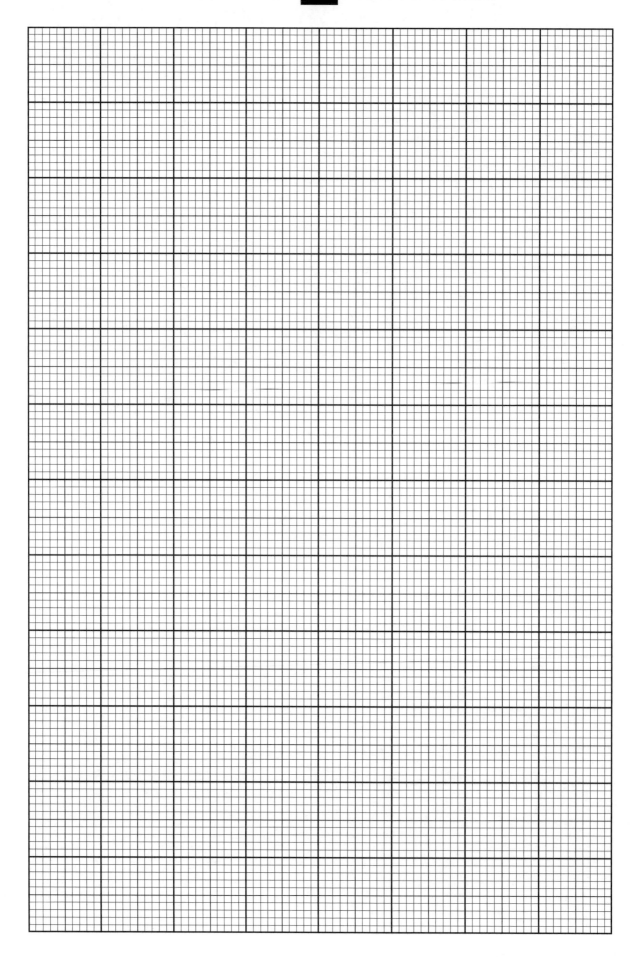

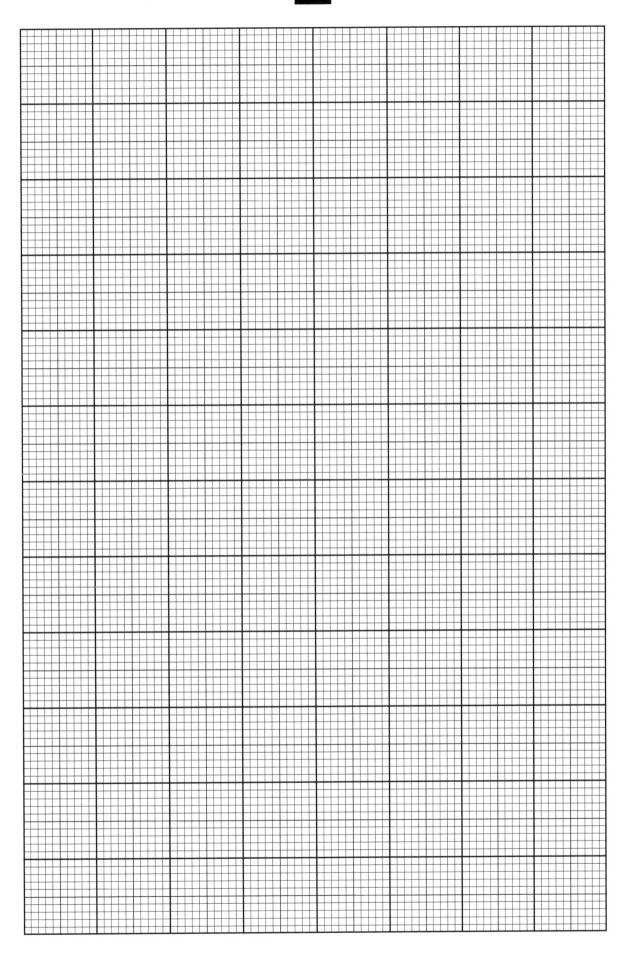

ADDITIONAL SPACE FOR ANSWERS AND ROUGH WORK

Additional graph for Question 11 (c)(iii)

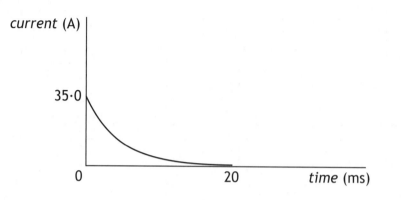

ADDITIONAL SPACE FOR ANSWERS AND ROUGH WORK

MARKS DO NOT WRITE IN THIS MARGIN

ADDITIONAL SPACE FOR ANSWERS AND ROUGH WORK

HIGHER

2016

National
Qualifications
2016

X757/76/02

**Physics
Section 1 — Questions**

TUESDAY, 24 MAY
9:00 AM – 11:30 AM

Instructions for the completion of Section 1 are given on *Page two* of your question and answer booklet X757/76/01.

Record your answers on the answer grid on *Page three* of your question and answer booklet.

Reference may be made to the Data Sheet on *Page two* of this booklet and to the Relationships Sheet X757/76/11.

Before leaving the examination room you must give your question and answer booklet to the Invigilator; if you do not, you may lose all the marks for this paper.

DATA SHEET

COMMON PHYSICAL QUANTITIES

Quantity	Symbol	Value	Quantity	Symbol	Value
Speed of light in vacuum	c	$3.00 \times 10^8\,\text{m s}^{-1}$	Planck's constant	h	$6.63 \times 10^{-34}\,\text{J s}$
Magnitude of the charge on an electron	e	$1.60 \times 10^{-19}\,\text{C}$	Mass of electron	m_e	$9.11 \times 10^{-31}\,\text{kg}$
Universal Constant of Gravitation	G	$6.67 \times 10^{-11}\,\text{m}^3\,\text{kg}^{-1}\,\text{s}^{-2}$	Mass of neutron	m_n	$1.675 \times 10^{-27}\,\text{kg}$
Gravitational acceleration on Earth	g	$9.8\,\text{m s}^{-2}$	Mass of proton	m_p	$1.673 \times 10^{-27}\,\text{kg}$
Hubble's constant	H_0	$2.3 \times 10^{-18}\,\text{s}^{-1}$			

REFRACTIVE INDICES

The refractive indices refer to sodium light of wavelength 589 nm and to substances at a temperature of 273 K.

Substance	Refractive index	Substance	Refractive index
Diamond	2·42	Water	1·33
Crown glass	1·50	Air	1·00

SPECTRAL LINES

Element	Wavelength/nm	Colour	Element	Wavelength/nm	Colour
Hydrogen	656	Red	Cadmium	644	Red
	486	Blue-green		509	Green
	434	Blue-violet		480	Blue
	410	Violet	Lasers		
	397	Ultraviolet	Element	Wavelength/nm	Colour
	389	Ultraviolet	Carbon dioxide	9550 } 10590	Infrared
Sodium	589	Yellow	Helium-neon	633	Red

PROPERTIES OF SELECTED MATERIALS

Substance	Density/kg m^{-3}	Melting Point/K	Boiling Point/K
Aluminium	2.70×10^3	933	2623
Copper	8.96×10^3	1357	2853
Ice	9.20×10^2	273	. . . .
Sea Water	1.02×10^3	264	377
Water	1.00×10^3	273	373
Air	1·29	. . . .	. . . .
Hydrogen	9.0×10^{-2}	14	20

The gas densities refer to a temperature of 273 K and a pressure of 1.01×10^5 Pa.

SECTION 1 — 20 marks
Attempt ALL questions

1. A car accelerates uniformly from rest. The car travels a distance of 60 m in 6·0 s. The acceleration of the car is

   A $0·83\,m\,s^{-2}$

   B $3·3\,m\,s^{-2}$

   C $5·0\,m\,s^{-2}$

   D $10\,m\,s^{-2}$

   E $20\,m\,s^{-2}$.

2. A ball is thrown vertically upwards and falls back to Earth.

   Neglecting air resistance, which velocity-time graph represents its motion?

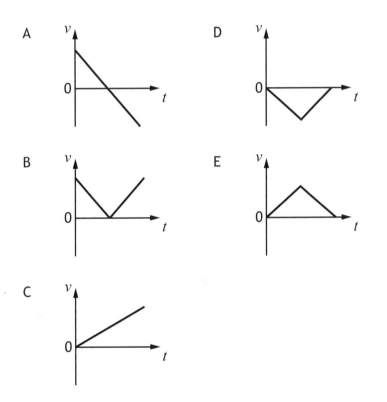

[Turn over

3. A block of wood slides with a constant velocity down a slope. The slope makes an angle of 30° with the horizontal as shown. The mass of the block is 2·0 kg.

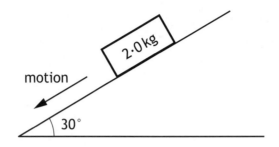

motion

2·0 kg

30°

The magnitude of the force of friction acting on the block is

A   1·0 N

B   1·7 N

C   9·8 N

D   17·0 N

E   19·6 N.

4. The graph shows the force which acts on an object over a time interval of 8·0 seconds.

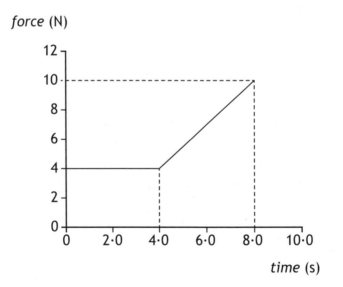

The momentum gained by the object during this 8·0 seconds is

A   12 kg m s^{-1}

B   32 kg m s^{-1}

C   44 kg m s^{-1}

D   52 kg m s^{-1}

E   72 kg m s^{-1}.

5. A planet orbits a star at a distance of $3 \cdot 0 \times 10^9$ m.

    The star exerts a gravitational force of $1 \cdot 6 \times 10^{27}$ N on the planet.

    The mass of the star is $6 \cdot 0 \times 10^{30}$ kg.

    The mass of the planet is

    A $2 \cdot 4 \times 10^{14}$ kg

    B $1 \cdot 2 \times 10^{16}$ kg

    C $3 \cdot 6 \times 10^{25}$ kg

    D $1 \cdot 6 \times 10^{26}$ kg

    E $2 \cdot 4 \times 10^{37}$ kg.

6. A car horn emits a sound with a constant frequency of 405 Hz.

    The car is travelling away from a student at $28 \cdot 0$ m s^{-1}.

    The speed of sound in air is 335 m s^{-1}.

    The frequency of the sound from the horn heard by the student is

    A   371 Hz

    B   374 Hz

    C   405 Hz

    D   439 Hz

    E   442 Hz.

[Turn over

7. The graphs show how the radiation per unit surface area, R, varies with the wavelength, λ, of the emitted radiation for two stars, P and Q.

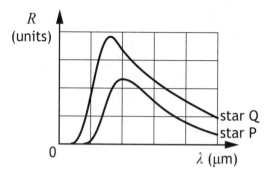

A student makes the following conclusions based on the information in the graph.

I   Star P is hotter than star Q.

II   Star P emits more radiation per unit surface area than star Q.

III The peak intensity of the radiation from star Q is at a shorter wavelength than that from star P.

Which of these statements is/are correct?

A   I only

B   II only

C   III only

D   I and II only

E   II and III only

8. One type of hadron consists of two down quarks and one up quark.

The charge on a down quark is $-\frac{1}{3}$.

The charge on an up quark is $+\frac{2}{3}$.

Which row in the table shows the charge and type for this hadron?

	charge	type of hadron
A	0	baryon
B	+1	baryon
C	−1	meson
D	0	meson
E	+1	meson

9. A student makes the following statements about sub-nuclear particles.

    I   The force mediating particles are bosons.

    II   Gluons are the mediating particles of the strong force.

    III Photons are the mediating particles of the electromagnetic force.

    Which of these statements is/are correct?

    A   I only

    B   II only

    C   I and II only

    D   II and III only

    E   I, II and III

10. The last two changes in a radioactive decay series are shown below.

    A Bismuth nucleus emits a beta particle and its product, a Polonium nucleus, emits an alpha particle.

    $$ {}^P_Q\text{Bi} \xrightarrow[\text{decay}]{\beta} {}^R_S\text{Po} \xrightarrow[\text{decay}]{\alpha} {}^{208}_{82}\text{Pb} $$

    Which numbers are represented by P, Q, R and S?

	P	Q	R	S
A	210	83	208	81
B	210	83	210	84
C	211	85	207	86
D	212	83	212	84
E	212	85	212	84

    [Turn over

11. The table below shows the threshold frequency of radiation for photoelectric emission for some metals.

Metal	Threshold frequency (Hz)
sodium	$4 \cdot 4 \times 10^{14}$
potassium	$5 \cdot 4 \times 10^{14}$
zinc	$6 \cdot 9 \times 10^{14}$

Radiation of frequency $6 \cdot 3 \times 10^{14}$ Hz is incident on the surface of each of the metals. Photoelectric emission occurs from

A   sodium only

B   zinc only

C   potassium only

D   sodium and potassium only

E   zinc and potassium only.

12. Radiation of frequency $9 \cdot 00 \times 10^{15}$ Hz is incident on a clean metal surface.

The maximum kinetic energy of a photoelectron ejected from this surface is $5 \cdot 70 \times 10^{-18}$ J. The work function of the metal is

A $2 \cdot 67 \times 10^{-19}$ J

B $5 \cdot 97 \times 10^{-18}$ J

C $1 \cdot 17 \times 10^{-17}$ J

D $2 \cdot 07 \times 10^{-2}$ J

E $9 \cdot 60 \times 10^{-1}$ J.

13. A ray of monochromatic light is incident on a grating as shown.

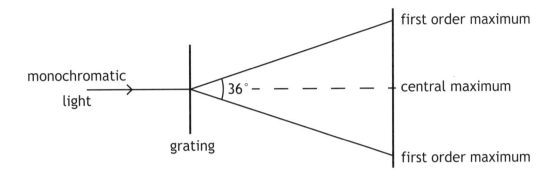

The wavelength of the light is 633 nm.

The separation of the slits on the grating is

A $1 \cdot 96 \times 10^{-7}$ m

B $1 \cdot 08 \times 10^{-6}$ m

C $2 \cdot 05 \times 10^{-6}$ m

D $2 \cdot 15 \times 10^{-6}$ m

E $4 \cdot 10 \times 10^{-6}$ m.

14. Light travels from **glass** into **air**.

Which row in the table shows what happens to the speed, frequency and wavelength of the light as it travels from glass into air?

	Speed	Frequency	Wavelength
A	decreases	stays constant	decreases
B	decreases	increases	stays constant
C	stays constant	increases	increases
D	increases	increases	stays constant
E	increases	stays constant	increases

15. The irradiance of light from a point source is $32 \, \mathrm{W \, m^{-2}}$ at a distance of $4 \cdot 0$ m from the source.

The irradiance of the light at a distance of 16 m from the source is

A $0 \cdot 125 \, \mathrm{W \, m^{-2}}$

B $0 \cdot 50 \, \mathrm{W \, m^{-2}}$

C $2 \cdot 0 \, \mathrm{W \, m^{-2}}$

D $8 \cdot 0 \, \mathrm{W \, m^{-2}}$

E $128 \, \mathrm{W \, m^{-2}}$.

16. Part of the energy level diagram for an atom is shown

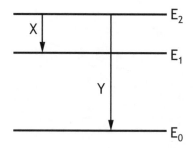

X and Y represent two possible electron transitions.

A student makes the following statements about transitions X and Y.

I   Transition Y produces photons of higher frequency than transition X

II   Transition X produces photons of longer wavelength than transition Y

III   When an electron is in the energy level E_0, the atom is ionised.

Which of the statements is/are correct?

A   I only

B   I and II only

C   I and III only

D   II and III only

E   I, II and III

17. The output of a signal generator is connected to the input of an oscilloscope.

The trace produced on the screen of the oscilloscope is shown.

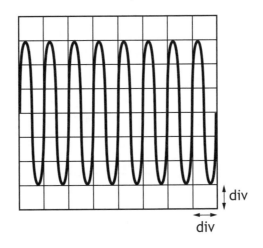

The timebase control of the oscilloscope is set at 2 ms/div.

The Y-gain control of the oscilloscope is set at 4 mV/div.

Which row in the table shows the frequency and peak voltage of the output of the signal generator?

	frequency (Hz)	peak voltage (mV)
A	0·5	12
B	0·5	6
C	250	6
D	500	12
E	500	24

[Turn over

18. A potential divider circuit is set up as shown.

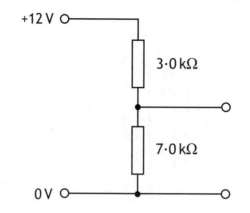

The potential difference across the 7·0 kΩ resistor is

A   3·6 V

B   4·0 V

C   5·1 V

D   8·4 V

E   9·0 V.

19. A circuit is set up as shown.

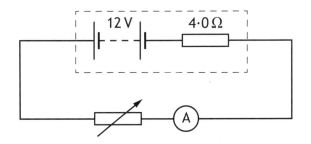

The resistance of the variable resistor is increased and corresponding readings on the ammeter are recorded.

Resistance (Ω)	2·0	4·0	6·0	8·0
Current (A)	2·0	1·5	1·2	1·0

These results show that as the resistance of the variable resistor increases the power dissipated in the variable resistor

A   increases

B   decreases

C   remains constant

D   decreases and then increases

E   increases and then decreases.

20. A 20 μF capacitor is connected to a 12 V d.c. supply.

The maximum charge stored on the capacitor is

A $1·4 \times 10^{-3}$ C

B $2·4 \times 10^{-4}$ C

C $1·2 \times 10^{-4}$ C

D $1·7 \times 10^{-6}$ C

E $6·0 \times 10^{-7}$ C.

[END OF SECTION 1. NOW ATTEMPT THE QUESTIONS IN SECTION 2
OF YOUR QUESTION AND ANSWER BOOKLET]

[BLANK PAGE]

DO NOT WRITE ON THIS PAGE

National
Qualifications
2016

X757/76/11

**Physics
Relationship Sheet**

TUESDAY, 24 MAY
9:00 AM – 11:30 AM

Relationships required for Physics Higher

$d = \bar{v}t$

$s = \bar{v}t$

$v = u + at$

$s = ut + \frac{1}{2}at^2$

$v^2 = u^2 + 2as$

$s = \frac{1}{2}(u + v)t$

$W = mg$

$F = ma$

$E_W = Fd$

$E_p = mgh$

$E_k = \frac{1}{2}mv^2$

$P = \frac{E}{t}$

$p = mv$

$Ft = mv - mu$

$F = G\frac{m_1 m_2}{r^2}$

$t' = \dfrac{t}{\sqrt{1 - \left(\frac{v}{c}\right)^2}}$

$l' = l\sqrt{1 - \left(\frac{v}{c}\right)^2}$

$f_o = f_s\left(\dfrac{v}{v \pm v_s}\right)$

$z = \dfrac{\lambda_{observed} - \lambda_{rest}}{\lambda_{rest}}$

$z = \dfrac{v}{c}$

$v = H_0 d$

$W = QV$

$E = mc^2$

$E = hf$

$E_k = hf - hf_0$

$E_2 - E_1 = hf$

$T = \dfrac{1}{f}$

$v = f\lambda$

$d \sin\theta = m\lambda$

$n = \dfrac{\sin\theta_1}{\sin\theta_2}$

$\dfrac{\sin\theta_1}{\sin\theta_2} = \dfrac{\lambda_1}{\lambda_2} = \dfrac{v_1}{v_2}$

$\sin\theta_c = \dfrac{1}{n}$

$I = \dfrac{k}{d^2}$

$I = \dfrac{P}{A}$

path difference $= m\lambda$ or $\left(m + \frac{1}{2}\right)\lambda$ where $m = 0, 1, 2 \ldots$

random uncertainty $= \dfrac{\text{max. value} - \text{min. value}}{\text{number of values}}$

$V_{peak} = \sqrt{2}V_{rms}$

$I_{peak} = \sqrt{2}I_{rms}$

$Q = It$

$V = IR$

$P = IV = I^2 R = \dfrac{V^2}{R}$

$R_T = R_1 + R_2 + \ldots$

$\dfrac{1}{R_T} = \dfrac{1}{R_1} + \dfrac{1}{R_2} + \ldots$

$E = V + Ir$

$V_1 = \left(\dfrac{R_1}{R_1 + R_2}\right)V_s$

$\dfrac{V_1}{V_2} = \dfrac{R_1}{R_2}$

$C = \dfrac{Q}{V}$

$E = \frac{1}{2}QV = \frac{1}{2}CV^2 = \frac{1}{2}\dfrac{Q^2}{C}$

Additional Relationships

Circle

circumference $= 2\pi r$

area $= \pi r^2$

Sphere

area $= 4\pi r^2$

volume $= \frac{4}{3}\pi r^3$

Trigonometry

$\sin \Theta = \dfrac{\text{opposite}}{\text{hypotenuse}}$

$\cos \Theta = \dfrac{\text{adjacent}}{\text{hypotenuse}}$

$\tan \Theta = \dfrac{\text{opposite}}{\text{adjacent}}$

$\sin^2 \Theta + \cos^2 \Theta = 1$

Electron Arrangements of Elements

Key

Atomic number
Symbol
Electron arrangement
Name

Transition Elements

Group 1	Group 2	(3)	(4)	(5)	(6)	(7)	(8)	(9)	(10)	(11)	(12)	Group 3 (13)	Group 4 (14)	Group 5 (15)	Group 6 (16)	Group 7 (17)	Group 0 (18)
1 **H** 1 Hydrogen	(2)																2 **He** 2 Helium
3 **Li** 2,1 Lithium	4 **Be** 2,2 Beryllium											5 **B** 2,3 Boron	6 **C** 2,4 Carbon	7 **N** 2,5 Nitrogen	8 **O** 2,6 Oxygen	9 **F** 2,7 Fluorine	10 **Ne** 2,8 Neon
11 **Na** 2,8,1 Sodium	12 **Mg** 2,8,2 Magnesium											13 **Al** 2,8,3 Aluminium	14 **Si** 2,8,4 Silicon	15 **P** 2,8,5 Phosphorus	16 **S** 2,8,6 Sulfur	17 **Cl** 2,8,7 Chlorine	18 **Ar** 2,8,8 Argon
19 **K** 2,8,8,1 Potassium	20 **Ca** 2,8,8,2 Calcium	21 **Sc** 2,8,9,2 Scandium	22 **Ti** 2,8,10,2 Titanium	23 **V** 2,8,11,2 Vanadium	24 **Cr** 2,8,13,1 Chromium	25 **Mn** 2,8,13,2 Manganese	26 **Fe** 2,8,14,2 Iron	27 **Co** 2,8,15,2 Cobalt	28 **Ni** 2,8,16,2 Nickel	29 **Cu** 2,8,18,1 Copper	30 **Zn** 2,8,18,2 Zinc	31 **Ga** 2,8,18,3 Gallium	32 **Ge** 2,8,18,4 Germanium	33 **As** 2,8,18,5 Arsenic	34 **Se** 2,8,18,6 Selenium	35 **Br** 2,8,18,7 Bromine	36 **Kr** 2,8,18,8 Krypton
37 **Rb** 2,8,18,8,1 Rubidium	38 **Sr** 2,8,18,8,2 Strontium	39 **Y** 2,8,18,9,2 Yttrium	40 **Zr** 2,8,18,10,2 Zirconium	41 **Nb** 2,8,18,12,1 Niobium	42 **Mo** 2,8,18,13,1 Molybdenum	43 **Tc** 2,8,18,13,2 Technetium	44 **Ru** 2,8,18,15,1 Ruthenium	45 **Rh** 2,8,18,16,1 Rhodium	46 **Pd** 2,8,18,18,0 Palladium	47 **Ag** 2,8,18,18,1 Silver	48 **Cd** 2,8,18,18,2 Cadmium	49 **In** 2,8,18,18,3 Indium	50 **Sn** 2,8,18,18,4 Tin	51 **Sb** 2,8,18,18,5 Antimony	52 **Te** 2,8,18,18,6 Tellurium	53 **I** 2,8,18,18,7 Iodine	54 **Xe** 2,8,18,18,8 Xenon
55 **Cs** 2,8,18,18,8,1 Caesium	56 **Ba** 2,8,18,18,8,2 Barium	57 **La** 2,8,18,18,9,2 Lanthanum	72 **Hf** 2,8,18,32,10,2 Hafnium	73 **Ta** 2,8,18,32,11,2 Tantalum	74 **W** 2,8,18,32,12,2 Tungsten	75 **Re** 2,8,18,32,13,2 Rhenium	76 **Os** 2,8,18,32,14,2 Osmium	77 **Ir** 2,8,18,32,15,2 Iridium	78 **Pt** 2,8,18,32,17,1 Platinum	79 **Au** 2,8,18,32,18,1 Gold	80 **Hg** 2,8,18,32,18,2 Mercury	81 **Tl** 2,8,18,32,18,3 Thallium	82 **Pb** 2,8,18,32,18,4 Lead	83 **Bi** 2,8,18,32,18,5 Bismuth	84 **Po** 2,8,18,32,18,6 Polonium	85 **At** 2,8,18,32,18,7 Astatine	86 **Rn** 2,8,18,32,18,8 Radon
87 **Fr** 2,8,18,32,18,8,1 Francium	88 **Ra** 2,8,18,32,18,8,2 Radium	89 **Ac** 2,8,18,32,18,9,2 Actinium	104 **Rf** 2,8,18,32,32,10,2 Rutherfordium	105 **Db** 2,8,18,32,32,11,2 Dubnium	106 **Sg** 2,8,18,32,32,12,2 Seaborgium	107 **Bh** 2,8,18,32,32,13,2 Bohrium	108 **Hs** 2,8,18,32,32,14,2 Hassium	109 **Mt** 2,8,18,32,32,15,2 Meitnerium	110 **Ds** 2,8,18,32,32,17,1 Darmstadtium	111 **Rg** 2,8,18,32,32,18,1 Roentgenium	112 **Cn** 2,8,18,32,32,18,2 Copernicium						

Lanthanides

57 **La** 2,8,18,18,9,2 Lanthanum	58 **Ce** 2,8,18,20,8,2 Cerium	59 **Pr** 2,8,18,21,8,2 Praseodymium	60 **Nd** 2,8,18,22,8,2 Neodymium	61 **Pm** 2,8,18,23,8,2 Promethium	62 **Sm** 2,8,18,24,8,2 Samarium	63 **Eu** 2,8,18,25,8,2 Europium	64 **Gd** 2,8,18,25,9,2 Gadolinium	65 **Tb** 2,8,18,27,8,2 Terbium	66 **Dy** 2,8,18,28,8,2 Dysprosium	67 **Ho** 2,8,18,29,8,2 Holmium	68 **Er** 2,8,18,30,8,2 Erbium	69 **Tm** 2,8,18,31,8,2 Thulium	70 **Yb** 2,8,18,32,8,2 Ytterbium	71 **Lu** 2,8,18,32,9,2 Lutetium

Actinides

89 **Ac** 2,8,18,32,18,9,2 Actinium	90 **Th** 2,8,18,32,18,10,2 Thorium	91 **Pa** 2,8,18,32,20,9,2 Protactinium	92 **U** 2,8,18,32,21,9,2 Uranium	93 **Np** 2,8,18,32,22,9,2 Neptunium	94 **Pu** 2,8,18,32,24,8,2 Plutonium	95 **Am** 2,8,18,32,25,8,2 Americium	96 **Cm** 2,8,18,32,25,9,2 Curium	97 **Bk** 2,8,18,32,27,8,2 Berkelium	98 **Cf** 2,8,18,32,28,8,2 Californium	99 **Es** 2,8,18,32,29,8,2 Einsteinium	100 **Fm** 2,8,18,32,30,8,2 Fermium	101 **Md** 2,8,18,32,31,8,2 Mendelevium	102 **No** 2,8,18,32,32,8,2 Nobelium	103 **Lr** 2,8,18,32,32,9,2 Lawrencium

H

National
Qualifications
2016

Mark

X757/76/01

Physics
Section 1 — Answer Grid
and Section 2

TUESDAY, 24 MAY

9:00 AM – 11:30 AM

Fill in these boxes and read what is printed below.

Full name of centre

Town

Forename(s)

Surname

Number of seat

Date of birth

Day | Month | Year

Scottish candidate number

Total marks — 130

SECTION 1 — 20 marks
Attempt ALL questions.
Instructions for the completion of Section 1 are given on *Page two*.

SECTION 2 — 110 marks
Attempt ALL questions.

Reference may be made to the Data Sheet on *Page two* of the question paper X757/76/02 and to the Relationships Sheet X757/76/11.

Care should be taken to give an appropriate number of significant figures in the final answers to calculations.

Write your answers clearly in the spaces provided in this booklet. Additional space for answers and rough work is provided at the end of this booklet. If you use this space you must clearly identify the question number you are attempting. Any rough work must be written in this booklet. You should score through your rough work when you have written your final copy.

Use **blue** or **black** ink.

Before leaving the examination room you must give this booklet to the Invigilator; if you do not, you may lose all the marks for this paper.

SECTION 1 — 20 marks

The questions for Section 1 are contained in the question paper X757/76/02.
Read these and record your answers on the answer grid on *Page three* opposite.
Use **blue** or **black** ink. Do NOT use gel pens or pencil.

1. The answer to each question is **either** A, B, C, D or E. Decide what your answer is, then fill in the appropriate bubble (see sample question below).

2. There is **only one correct** answer to each question.

3. Any rough work must be written in the additional space for answers and rough work at the end of this booklet.

Sample Question

The energy unit measured by the electricity meter in your home is the:

    A   ampere

    B   kilowatt-hour

    C   watt

    D   coulomb

    E   volt.

The correct answer is **B**—kilowatt-hour.   The answer **B** bubble has been clearly filled in (see below).

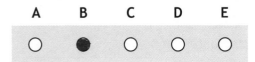

Changing an answer

If you decide to change your answer, cancel your first answer by putting a cross through it (see below) and fill in the answer you want. The answer below has been changed to **D**.

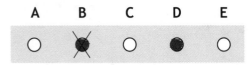

If you then decide to change back to an answer you have already scored out, put a tick (✓) to the **right** of the answer you want, as shown below:

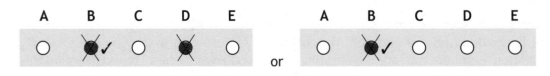

SECTION 1 — Answer Grid

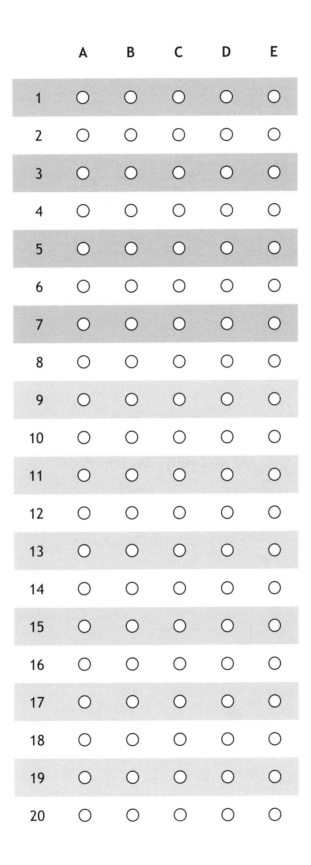

[Turn over

[BLANK PAGE]

DO NOT WRITE ON THIS PAGE

[Turn over for SECTION 2 on *Page six*

DO NOT WRITE ON THIS PAGE

MARKS | DO NOT WRITE IN THIS MARGIN

SECTION 2 — 110 marks

Attempt ALL questions

1.

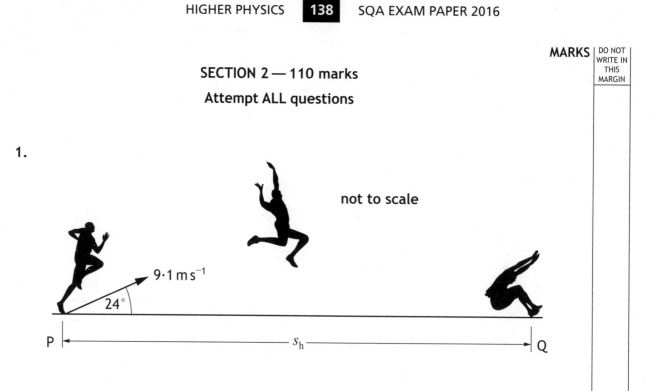

not to scale

An athlete takes part in a long jump competition. The athlete takes off from point P with an initial velocity of $9\cdot1\,\mathrm{m\,s^{-1}}$ at an angle of 24° to the horizontal and lands at point Q.

(a) Calculate:

   (i) the vertical component of the initial velocity of the athlete;   1

   Space for working and answer

   (ii) the horizontal component of the initial velocity of the athlete.   1

   Space for working and answer

MARKS | DO NOT WRITE IN THIS MARGIN

1. (continued)

(b) Show that the time taken for the athlete to travel from P to Q is 0·76 s.   **2**

Space for working and answer

(c) Calculate the horizontal displacement s_h between points P and Q.   **3**

Space for working and answer

(d) The graph shows how the horizontal displacement of the athlete varies with time for this jump when air resistance is ignored.

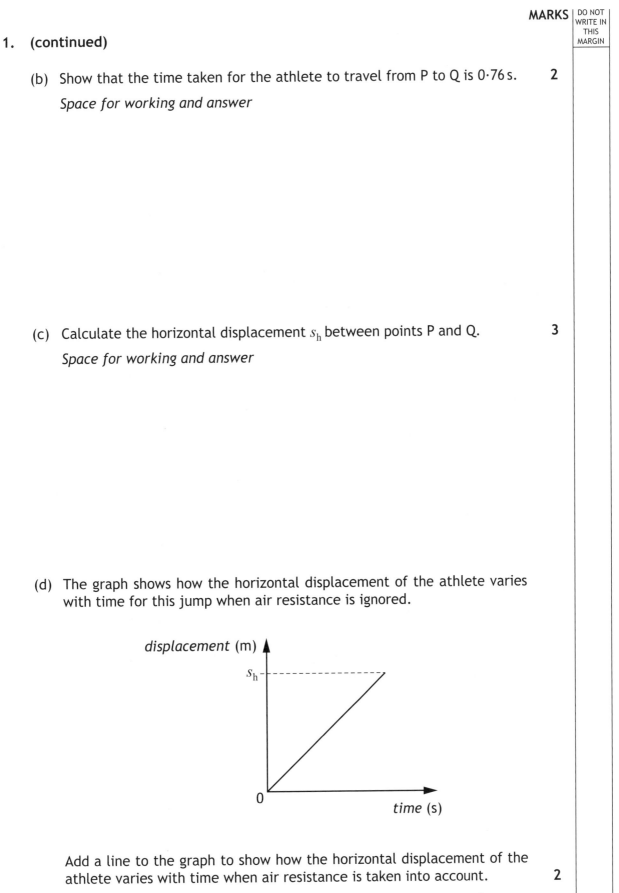

Add a line to the graph to show how the horizontal displacement of the athlete varies with time when air resistance is taken into account.   **2**

(An additional graph, if required can be found on *Page thirty-eight*.)

[Turn over

MARKS | DO NOT WRITE IN THIS MARGIN

2. A student uses the apparatus shown to investigate the force of friction between the wheels of a toy car and a carpet.

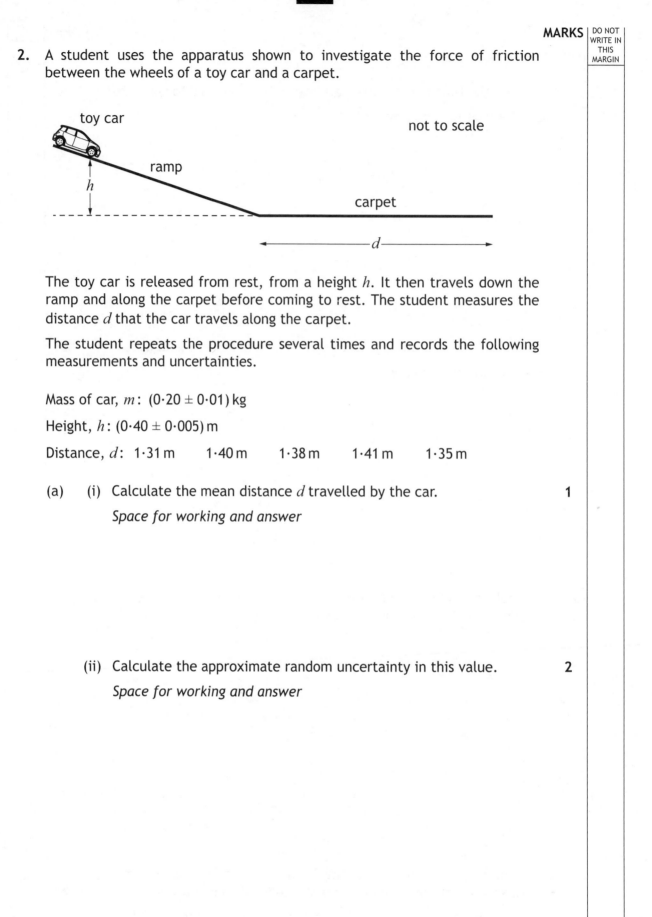

toy car

ramp

h

not to scale

carpet

d

The toy car is released from rest, from a height h. It then travels down the ramp and along the carpet before coming to rest. The student measures the distance d that the car travels along the carpet.

The student repeats the procedure several times and records the following measurements and uncertainties.

Mass of car, m: $(0.20 \pm 0.01)\,\text{kg}$

Height, h: $(0.40 \pm 0.005)\,\text{m}$

Distance, d: 1.31 m   1.40 m   1.38 m   1.41 m   1.35 m

(a)　(i)　Calculate the mean distance d travelled by the car. **1**

Space for working and answer

(ii)　Calculate the approximate random uncertainty in this value. **2**

Space for working and answer

MARKS | DO NOT WRITE IN THIS MARGIN

2. (continued)

(b) Determine which of the quantities; mass m, height h or mean distance d, has the largest percentage uncertainty.

You must justify your answer by calculation.   **4**

Space for working and answer

(c) (i) Calculate the potential energy of the toy car at height h.

An uncertainty in this value is not required.   **3**

Space for working and answer

[Turn over

MARKS | DO NOT WRITE IN THIS MARGIN

2. (c) (continued)

(ii) Calculate the average force of friction acting between the toy car and carpet, as the car comes to rest.

An uncertainty in this value is not required.   **3**

Space for working and answer

(iii) State one assumption you have made in (c) (ii).   **1**

[Turn over for next question

DO NOT WRITE ON THIS PAGE

MARKS DO NOT WRITE IN THIS MARGIN

3. The following apparatus is set up to investigate the law of conservation of linear momentum.

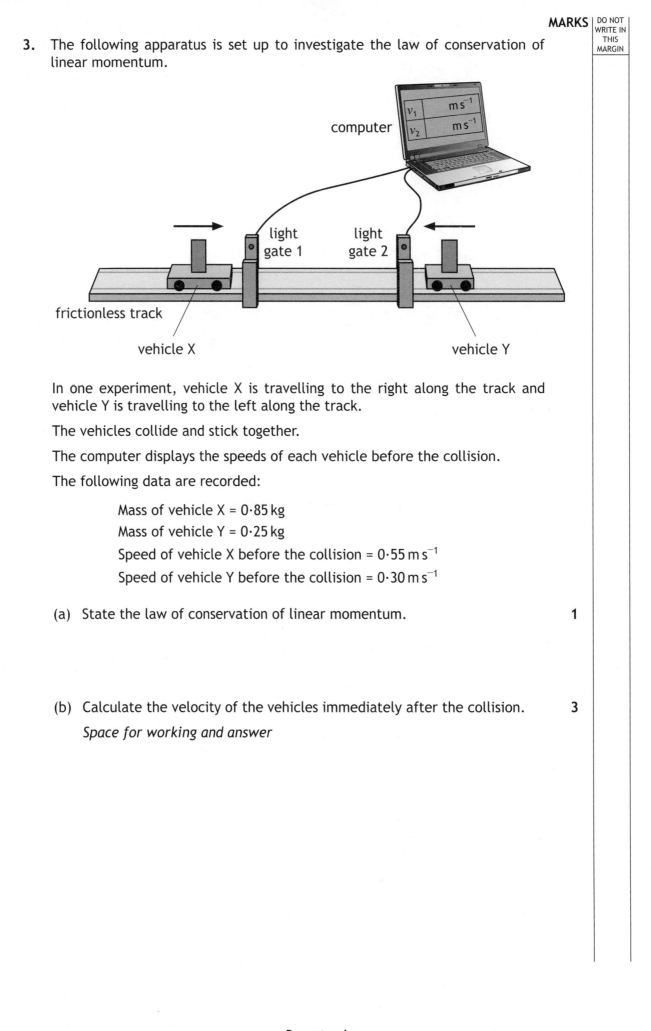

In one experiment, vehicle X is travelling to the right along the track and vehicle Y is travelling to the left along the track.

The vehicles collide and stick together.

The computer displays the speeds of each vehicle before the collision.

The following data are recorded:

Mass of vehicle X = 0·85 kg
Mass of vehicle Y = 0·25 kg
Speed of vehicle X before the collision = 0·55 m s^{-1}
Speed of vehicle Y before the collision = 0·30 m s^{-1}

(a) State the law of conservation of linear momentum. 1

(b) Calculate the velocity of the vehicles immediately after the collision. 3

Space for working and answer

MARKS | DO NOT WRITE IN THIS MARGIN

3. **(continued)**

(c) Show by calculation that the collision is inelastic.   **4**

Space for working and answer

[Turn over

MARKS | DO NOT WRITE IN THIS MARGIN

4. Two physics students are in an airport building on their way to visit CERN.

(a) The first student steps onto a moving walkway, which is travelling at $0.83\,\mathrm{m\,s^{-1}}$ relative to the building. This student walks along the walkway at a speed of $1.20\,\mathrm{m\,s^{-1}}$ relative to the walkway.

The second student walks alongside the walkway at a speed of $1.80\,\mathrm{m\,s^{-1}}$ relative to the building.

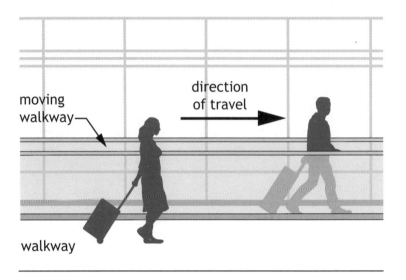

moving walkway

direction of travel

walkway

Determine the speed of the first student relative to the second student.   **2**

Space for working and answer

MARKS | DO NOT WRITE IN THIS MARGIN

4. (continued)

(b) On the plane, the students discuss the possibility of travelling at relativistic speeds.

  (i) The students consider the plane travelling at $0·8c$ relative to a stationary observer. The plane emits a beam of light towards the observer.

  State the speed of the emitted light as measured by the observer.

  Justify your answer.   **2**

  (ii) According to the manufacturer, the length of the plane is 71 m.

  Calculate the length of the plane travelling at $0·8c$ as measured by the stationary observer.   **3**

  Space for working and answer

  (iii) One of the students states that the clocks on board the plane will run slower when the plane is travelling at relativistic speeds.

  Explain whether or not this statement is correct.   **1**

[Turn over

5. (a) A student is using an elastic band to model the expansion of the Universe.

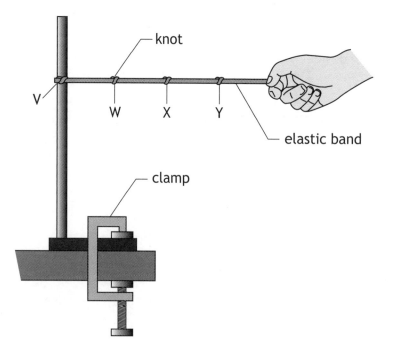

One end of the band is fixed in a clamp stand at V. Knots are tied in the band to represent galaxies. The knots are at regular intervals of 0·10 m, at points W, X and Y as shown.

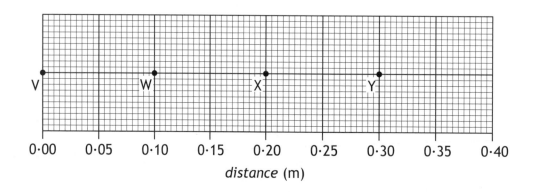

The other end of the elastic band is pulled slowly for 2·5 seconds, so that the band stretches. The knots are now in the positions shown below.

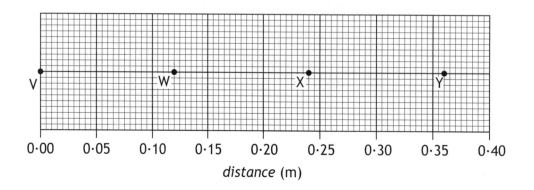

MARKS | DO NOT WRITE IN THIS MARGIN

5.　(a)　(continued)

(i)　Complete the table to show the average speeds of the knots X and Y.　　2

Knot	Average speed (m s^{-1})
W	0·008
X	
Y	

Space for working

(ii)　Explain why this model is a good simulation of the expansion of the Universe.　　1

[Turn over

MARKS | DO NOT WRITE IN THIS MARGIN

5. (continued)

(b) When viewed from the Earth, the continuous emission spectrum from the Sun has a number of dark lines. One of these lines is at a wavelength of 656 nm.

— 656 nm

In the spectrum of light from a distant galaxy, the corresponding dark line is observed at 667 nm.

Calculate the redshift of the light from the distant galaxy.   3

Space for working and answer

MARKS | DO NOT WRITE IN THIS MARGIN

6. A website states *"Atoms are like tiny solar systems with electrons orbiting a nucleus like the planets orbit the Sun"*.

   Use your knowledge of physics to comment on this statement. 3

[Turn over

MARKS | DO NOT WRITE IN THIS MARGIN

7. An experiment is set up to investigate the behaviour of electrons in electric fields.

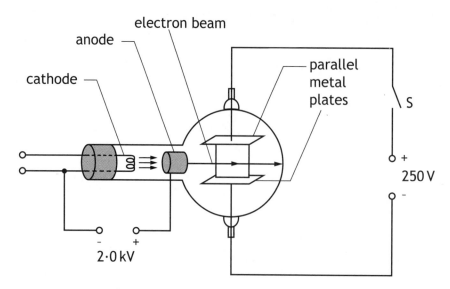

(a) Electrons are accelerated from rest between the cathode and the anode by a potential difference of 2·0 kV.

Calculate the kinetic energy gained by each electron as it reaches the anode.

Space for working and answer

3

(b) The electrons then pass between the two parallel metal plates.

The electron beam current is 8·0 mA.

Determine the number of electrons passing between the metal plates in one minute.

Space for working and answer

4

MARKS

DO NOT WRITE IN THIS MARGIN

7. (continued)

(c) The switch S is now closed.

The potential difference between the metal plates is 250 V.

The path of the electron beam between the metal plates is shown.

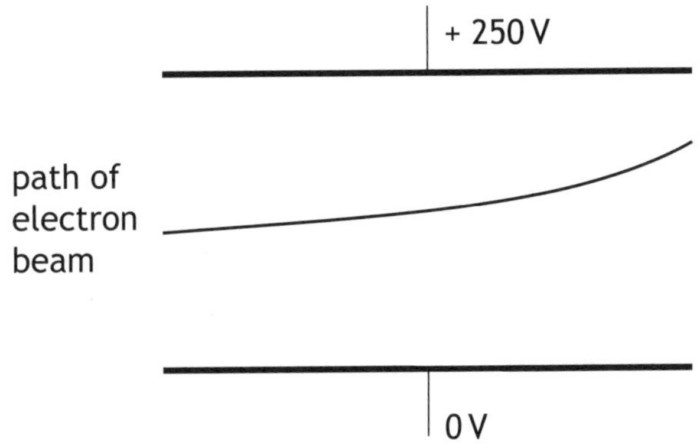

+ 250 V

path of
electron
beam

0 V

Complete the diagram to show the electric field pattern between the two
metal plates.

(An additional diagram, if required, can be found on *Page thirty-eight*.)

1

[Turn over

MARKS | DO NOT WRITE IN THIS MARGIN

8. The diagram shows part of an experimental fusion reactor.

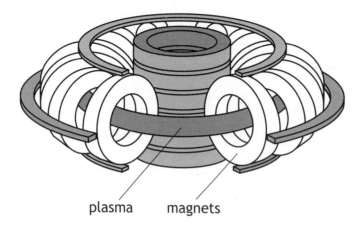

plasma   magnets

The following statement represents a reaction that takes place inside the reactor.

$$^2_1H + ^3_1H \rightarrow ^4_2He + ^1_0n$$

The masses of the particles involved in the reaction are shown in the table.

Particle	Mass (kg)
^2_1H	$3 \cdot 3436 \times 10^{-27}$
^3_1H	$5 \cdot 0083 \times 10^{-27}$
^4_2He	$6 \cdot 6465 \times 10^{-27}$
^1_0n	$1 \cdot 6749 \times 10^{-27}$

(a) Explain why energy is released in this reaction. 1

(b) Calculate the energy released in this reaction. 4

Space for working and answer

MARKS | DO NOT WRITE IN THIS MARGIN

8. (continued)

(c) Magnetic fields are used to contain the plasma inside the fusion reactor.

Explain why it is necessary to use a magnetic field to contain the plasma. **1**

(d) The plasma consists of charged particles. A positively charged particle enters a region of the magnetic field as shown.

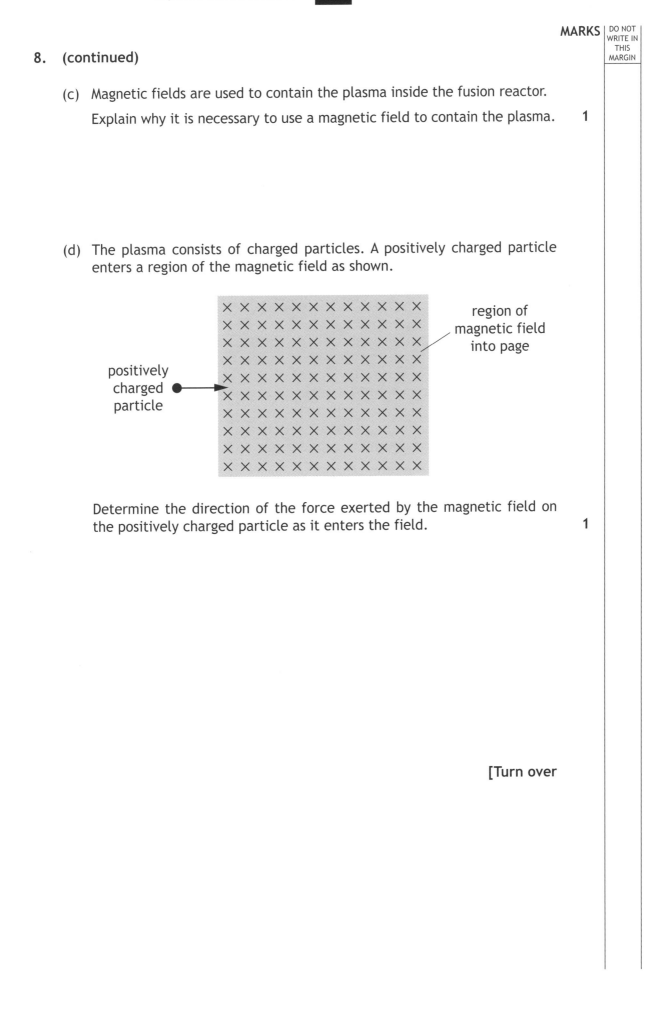

Determine the direction of the force exerted by the magnetic field on the positively charged particle as it enters the field. **1**

[Turn over

MARKS | DO NOT WRITE IN THIS MARGIN

9. A student carries out an experiment to measure the wavelength of microwave radiation. Microwaves pass through two gaps between metal plates as shown.

As the detector is moved from A to B, a series of maxima and minima are detected.

(a) The microwaves passing through the gaps are coherent.

State what is meant by the term *coherent*. 1

(b) Explain, in terms of waves, how a maximum is produced. 1

(c) The measurements of the distance from each gap to the second order maximum are shown in the diagram above.

Calculate the wavelength of the microwaves. 3

Space for working and answer

MARKS | DO NOT WRITE IN THIS MARGIN

9. **(continued)**

(d) The distance separating the two gaps is now increased.

State what happens to the path difference to the second order maximum.

Justify your answer.

2

[Turn over

MARKS | DO NOT WRITE IN THIS MARGIN

10. Retroflective materials reflect light to enhance the visibility of clothing.

One type of retroflective material is made from small glass spheres partially embedded in a silver-coloured surface that reflects light.

A ray of monochromatic light follows the path shown as it enters one of the glass spheres.

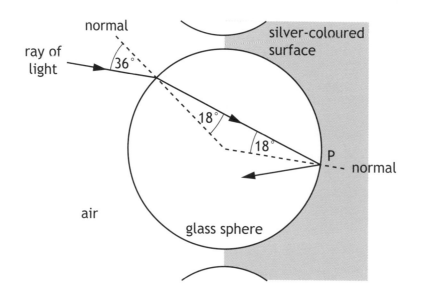

(a) Calculate the refractive index of the glass for this light.   3

Space for working and answer

MARKS | DO NOT WRITE IN THIS MARGIN

10. **(continued)**

(b) Calculate the critical angle for this light in the glass.   **3**

Space for working and answer

(c) The light is reflected at point P.

Complete the diagram below to show the path of the ray as it passes through the sphere and emerges into the air.   **1**

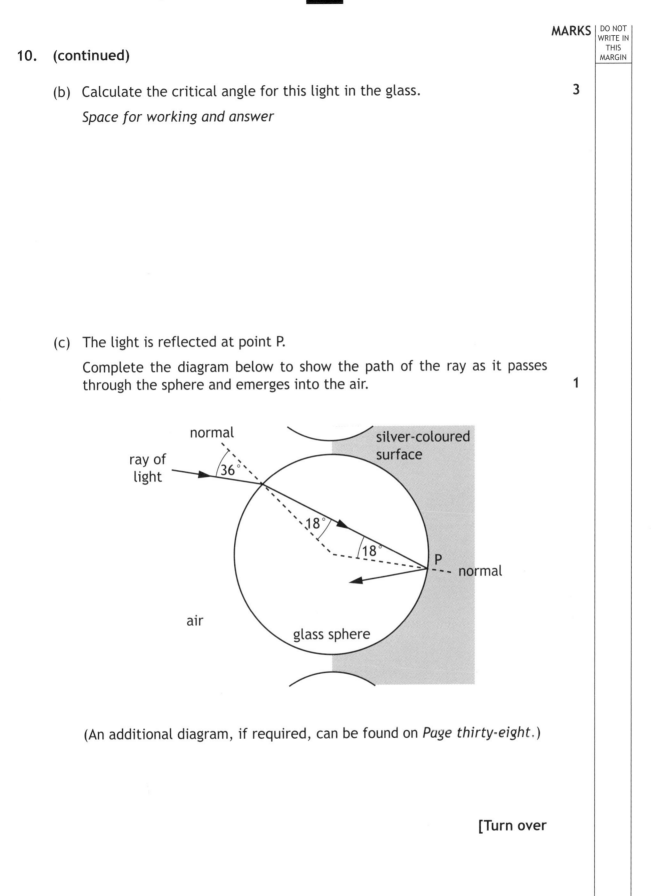

(An additional diagram, if required, can be found on *Page thirty-eight.*)

[Turn over

MARKS | DO NOT WRITE IN THIS MARGIN

11. A student is describing how the following circuit works.

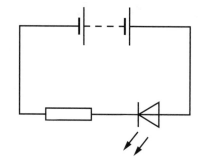

The student states:

"The electricity comes out of the battery with energy and flows through the resistor using up some of the energy, it then goes through the LED and the rest of the energy is changed into light waves."

Use your knowledge of physics to comment on this statement.

3

MARKS | DO NOT WRITE IN THIS MARGIN

12. A technician sets up a circuit as shown, using a car battery and two identical lamps.

The battery has an e.m.f. of 12·8 V and an internal resistance of 0·10 Ω.

(a) Switch S is open. The reading on the ammeter is 1·80 A.

   (i) Determine the reading on the voltmeter.

   Space for working and answer

   4

   (ii) Switch S is now closed.

   State the effect this has on the reading on the voltmeter.

   Justify your answer.

   3

MARKS | DO NOT WRITE IN THIS MARGIN

12. (continued)

(b) Some cars use LEDs in place of filament lamps.

An LED is made from semiconductor material that has been doped with impurities to create a p-n junction.

The diagram represents the band structure of an LED.

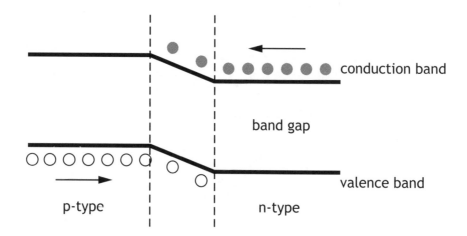

(i) A voltage is applied across an LED so that it is forward biased and emits light.

Using **band theory,** explain how the LED emits light.   3

MARKS | DO NOT WRITE IN THIS MARGIN

12. (b) (continued)

(ii) The energy gap between the valence band and conduction band is known as the band gap.

The band gap for the LED is $3 \cdot 03 \times 10^{-19}$ J

(A) Calculate the wavelength of the light emitted by the LED.   **4**

Space for working and answer

(B) Determine the colour of the light emitted by the LED.   **1**

[Turn over

MARKS | DO NOT WRITE IN THIS MARGIN

13. A technician sets up a circuit as shown.

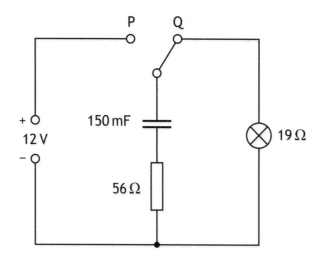

The power supply has negligible internal resistance.

(a) The capacitor is initially uncharged.

The switch is moved to position P and the capacitor charges.

(i) State the potential difference across the capacitor when it is fully charged.   **1**

(ii) Calculate the maximum energy stored by the capacitor.   **3**

Space for working and answer

MARKS | DO NOT WRITE IN THIS MARGIN

13. **(continued)**

(b) The switch is now moved back to position Q.

Determine the maximum discharge current in the circuit.   **3**

Space for working and answer

(c) The technician replaces the 150 mF capacitor with a capacitor of capacitance 47 mF.

The switch is moved to position P and the capacitor is fully charged.

The switch is now moved to position Q.

State the effect that this change has on the time the lamp stays lit.

You must justify your answer.   **2**

[Turn over for next question

MARKS | DO NOT WRITE IN THIS MARGIN

14. A student investigates the factors affecting the frequency of sound produced by a vibrating guitar string.

The guitar string is stretched over two supports and is made to vibrate as shown.

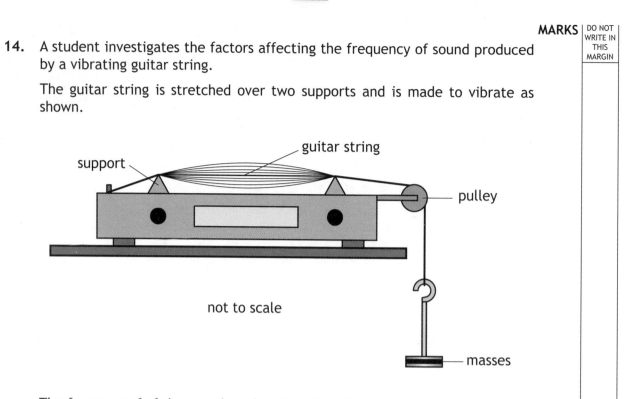

The frequency f of the sound produced by the vibrating string is given by the relationship

$$f = \frac{1}{2L}\sqrt{\frac{T}{\mu}}$$

where T is the tension in the string

L is the distance between the supports

μ is the mass per unit length of the string.

(a) The tension in the string is $49\cdot0\,\text{N}$ and the mass per unit length of the string is $4\cdot00 \times 10^{-4}\,\text{kg m}^{-1}$.

The distance between the supports is $0\cdot550\,\text{m}$.

Calculate the frequency f of the sound produced.   2

Space for working and answer

14. (continued)

(b) The guitar string in part (a) is replaced by a different guitar string.

A student varies the tension T and measures the frequency f of the sound produced by the new guitar string.

The student records the following information.

T (N)	$\sqrt{T}$ (N$^{\frac{1}{2}}$)	f (Hz)
10	3·2	162
15	3·9	190
20	4·5	220
25	5·0	254
30	5·5	273

(i) Using the square-ruled paper on *Page thirty-six*, draw a graph of f against $\sqrt{T}$ 3

(ii) Use your graph to determine the frequency of the sound produced when the tension in the guitar string is 22 N.   1

[END OF QUESTION PAPER]

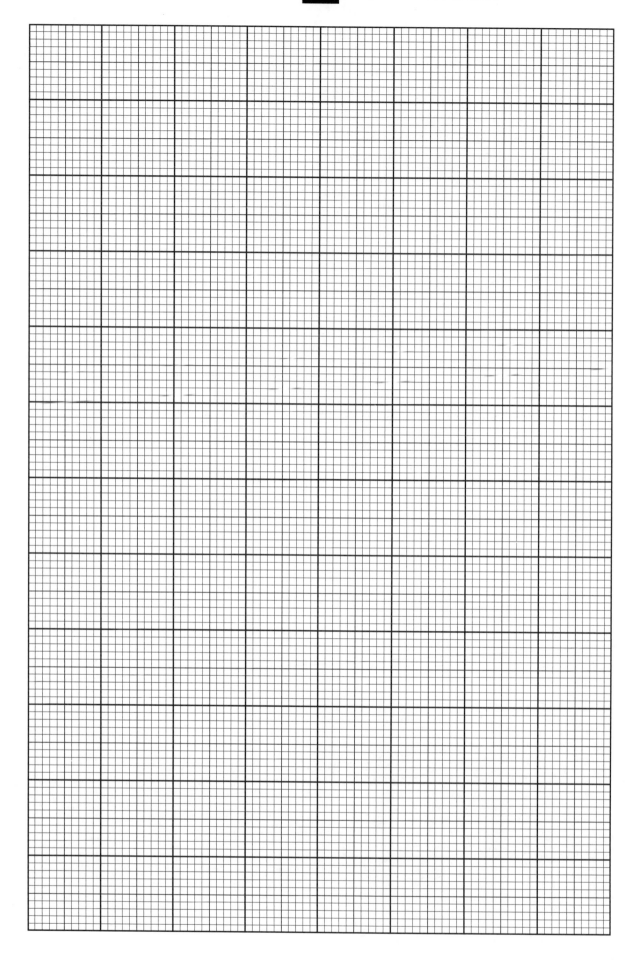

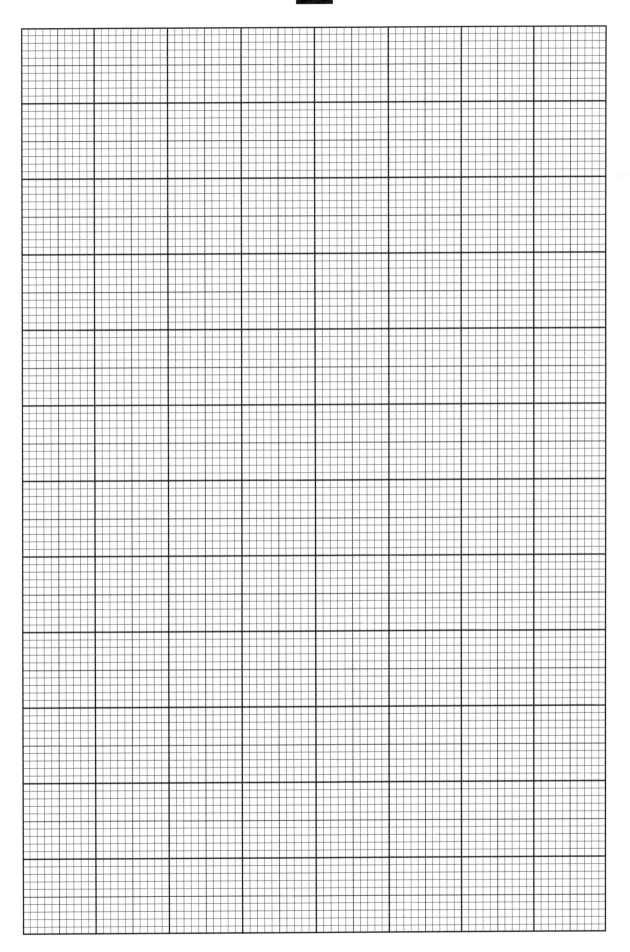

ADDITIONAL SPACE FOR ANSWERS AND ROUGH WORK

Question 1 (d)

Question 7 (c)

Question 10 (c)

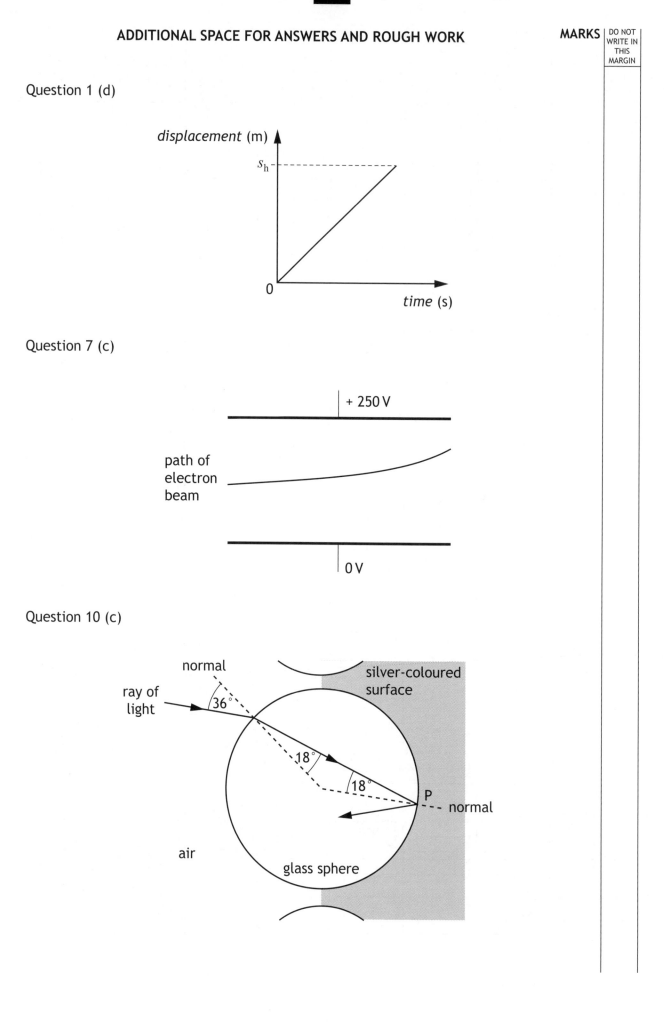

ADDITIONAL SPACE FOR ANSWERS AND ROUGH WORK

MARKS | DO NOT WRITE IN THIS MARGIN

ADDITIONAL SPACE FOR ANSWERS AND ROUGH WORK

HIGHER

Answers

SQA HIGHER PHYSICS 2016

SECTION 1

Question	Answer	Mark
1.	E	1
2.	A	1
3.	D	1
4.	B	1
5.	A	1
6.	B	1
7.	E	1
8.	C	1
9.	C	1
10.	A	1
11.	E	1
12.	A	1
13.	C	1
14.	A	1
15.	E	1
16.	D	1
17.	E	1
18.	C	1
19.	A	1
20.	B	1

SECTION 2

Question			Answer	Max mark
1.	(a)	(i)	(Initial horizontal component = $v\cos\theta$ = 50 cos35) = 41 ms^{-1} (1)	1
		(ii)	(Initial vertical component = $v\sin\theta$ = 50 sin 35) = 29 ms^{-1} (1)	1
	(b)		$v = u + at$ (1) $v = 29 - 9\cdot8t$ (1) $t = (0 - 29)/-9\cdot8$ $= 2\cdot96$ (s) $t_{TOTAL} = 5\cdot92$ (s) (1) $d_h = v_h t$ $= 41 \times 5\cdot92$ (1) (= 240 m)	4
2.	(a)	(i)	Component of weight down slope $= mg\sin\theta$ (1) $= 220 \times 9\cdot8 \times \sin 3\cdot2°$ (1) (=120 N)	2

Question			Answer	Max mark
		(ii)	Unbalanced Force = 230 – (120+48) = 62 N (1) $F = ma$ (1) 62 = 220 × a (1) $a = 0\cdot28$ m s^{-2} (1)	4
		(iii)	As angle (of slope) increases $mg\sin\theta$ increases (1) When $mg\sin\theta \gtrsim$ engine force – friction, the vehicle cannot move up the slope (1)	2
	(b)	(i)	lost volts = Ir (1) = 22 × 0·52 (1) (= 11 V)	2
		(ii)	p.d. = 48 – 11 = 37 V (1) $P = I\,V$ (1) = 22 × 37 (1) = 810 W (1)	4
	(c)		terminal potential difference decreases (1) current increases (1) lost volts increases (1)	3
3.			estimate of masses (500 kg < car mass < 3000 kg) (1) estimate of speed (10 ms^{-1} < speed < 70 ms^{-1}) (1) $E_k = 1/2\ mv^2$ (1) Final answer and unit (1)	4
4.	(a)		$d = vt$ (1) $d = (3 \times 10^8 \times 0\cdot995) \times 2\cdot2 \times 10^6$ (1) $d = 660$ m	2
	(b)		$t' = \dfrac{t}{\sqrt{1 - \left(\dfrac{v}{c}\right)^2}}$ (1) $t' = \dfrac{2\cdot2 \times 10^{-6}}{\sqrt{1 - \left(\dfrac{0\cdot995}{1}\right)^2}}$ (1) $= 2\cdot2 \times 10^{-5}$ s (1)	3
	(c)		For an observer on Earth's frame of reference the mean life of the muon is much greater **OR** The distance in the muon frame of reference is shorter	1

Question			Answer	Max mark
5.			This open-ended question requires comment on the **suitability of the design of the bicycle helmet**. Candidate responses are expected to make judgements on its suitability, on the basis of relevant physics ideas/concepts which might include one or more of: 'crumple zone'; impulse; energy being absorbed; air circulation and aerodynamics; or other relevant ideas/concepts.	
			3 marks: The candidate has demonstrated a **good** conceptual understanding of the physics involved, providing a logically correct response to the problem/ situation presented. This type of response might include a statement of principle(s) involved, a relationship or equation, and the application of these to respond to the problem/ situation. This does not mean the answer has to be what might be termed an 'excellent' answer or a 'complete' one. In response to this question, a **good** understanding might be demonstrated by a candidate response that: • makes a judgement on suitability based on one relevant physics idea/concept, in a **detailed/ developed** response that is **correct or largely correct** (any weaknesses are minor and do not detract from the overall response), **OR** • makes judgement(s) on suitability based on a range of relevant physics ideas/concepts, in a response that is **correct or largely correct** (any weaknesses are minor and do not detract from the overall response), **OR** • otherwise demonstrates a good understanding of the physics involved.	

Question			Answer	Max mark
			2 marks: The candidate has demonstrated a **reasonable** understanding of the physics involved, showing that the problem/situation is understood. This type of response might make some statement(s) that is/ are relevant to the problem/ situation, for example, a statement of relevant principle(s) or identification of a relevant relationship or equation. In response to this question, a **reasonable** understanding might be demonstrated by a candidate response that: • makes a judgement on suitability based on one or more relevant physics idea(s)/concept(s), in a response that is **largely correct** but has **weaknesses** which detract to a small extent from the overall response, **OR** • otherwise demonstrates a reasonable understanding of the physics involved.	
			1 mark: The candidate has demonstrated a **limited** understanding of the physics involved, showing that a little of the physics that is relevant to the problem/situation is understood. The candidate has made some statement(s) that is/are relevant to the problem/situation. In response to this question, a **limited** understanding might be demonstrated by a candidate response that: • makes a judgement on suitability based on one or more relevant physics idea(s)/concept(s), in a response that has **weaknesses** which detract to a large extent from the overall response, **OR** • otherwise demonstrates a limited understanding of the physics involved.	

Question			Answer		Max mark
			0 marks: The candidate has demonstrated **no** understanding of the physics that is relevant to the problem/situation. The candidate has made no statement(s) that is/are relevant to the problem/situation. Where the candidate has only demonstrated knowledge and understanding of physics **that is not relevant to the problem/situation presented**, 0 marks should be awarded.		
6.	(a)	(i)	The star is moving away from the Earth (1) Plus any one point from the following for 1 mark: • The apparent wavelength of the hydrogen spectra from the star has increased • The apparent frequency of the hydrogen spectra from the star is less than the actual frequency on Earth • The frequency of the light from the star has shifted towards the red end of the spectrum • Light from the star is experiencing a Doppler shift.		2
		(ii)	$z = \dfrac{(\lambda_{obs} - \lambda_{rest})}{\lambda_{rest}}$ (1) $z = \dfrac{(676 \times 10^{-9} - 656 \times 10^{-9})}{656 \times 10^{-9}}$ (1) $z = 0 \cdot 03$ (1) $z = \dfrac{v}{c}$ (1) $v = 0 \cdot 03c$ $v = 9 \times 10^{6} \text{ m s}^{-1}$ (1)		5
	(b)		$v = H_0 d$ (1) $d = \dfrac{v}{H_0}$ $d = \dfrac{1 \cdot 2 \times 10^{7}}{2 \cdot 3 \times 10^{-18}}$ (1) $d = 5 \cdot 2 \times 10^{24} \text{ m}$		2

Question			Answer		Max mark
	(c)		This open-ended question requires comment on the **suitability of the expanding balloon model to explain the expansion of the universe.** Candidate responses are expected to make judgements on its suitability, on the basis of relevant physics ideas/concepts which might include one or more of: that distances between the dots increase similarly as the distances between the galaxies; it is the 2-dimensional surface that is being compared to 3-dimensional space – so centre of balloon has no physical analogue; galaxies themselves do not expand – they are bound by gravitation; or other relevant ideas/concepts.		
			3 marks: The candidate has demonstrated a **good** conceptual understanding of the physics involved, providing a logically correct response to the problem/situation presented. This type of response might include a statement of principle(s) involved, a relationship or equation, and the application of these to respond to the problem/situation. This does not mean the answer has to be what might be termed an 'excellent' answer or a 'complete' one. In response to this question, a **good** understanding might be demonstrated by a candidate response that: • makes a judgement on suitability based on one relevant physics idea/concept, in a **detailed/developed** response that is **correct or largely correct** (any weaknesses are minor and do not detract from the overall response), **OR** • makes judgement(s) on suitability based on a range of relevant physics ideas/concepts, in a response that is **correct or largely correct** (any weaknesses are minor and do not detract from the overall response), **OR** • otherwise demonstrates a good understanding of the physics involved.		

Question			Answer	Max mark
			2 marks: The candidate has demonstrated a **reasonable** understanding of the physics involved, showing that the problem/situation is understood. This type of response might make some statement(s) that is/are relevant to the problem/situation, for example, a statement of relevant principle(s) or identification of a relevant relationship or equation. In response to this question, a **reasonable** understanding might be demonstrated by a candidate response that: • makes a judgement on suitability based on one or more relevant physics idea(s)/concept(s), in a response that is **largely correct** but has **weaknesses** which detract to a small extent from the overall response, **OR** • otherwise demonstrates a reasonable understanding of the physics involved.	
			1 mark: The candidate has demonstrated a **limited** understanding of the physics involved, showing that a little of the physics that is relevant to the problem/situation is understood. The candidate has made some statement(s) that is/are relevant to the problem/situation. In response to this question, a **limited** understanding might be demonstrated by a candidate response that: • makes a judgement on suitability based on one or more relevant physics idea(s)/concept(s), in a response that has **weaknesses** which detract to a large extent from the overall response, **OR** • otherwise demonstrates a limited understanding of the physics involved.	

Question			Answer	Max mark
			0 marks: The candidate has demonstrated **no** understanding of the physics that is relevant to the problem/situation. The candidate has made no statement(s) that is/are relevant to the problem/situation. Where the candidate has only demonstrated knowledge and understanding of physics **that is not relevant to the problem/situation presented**, 0 marks should be awarded.	
7.	(a)	(i)	A = 2u + 1d	1
			B = 1u + 2d	1
		(ii)	gluon	1
	(b)		beta decay	1
8.	(a)	(i)	$W = QV$ or $E_w = QV$ (1) $E_w = 1 \cdot 6 \times 10^{-19} \times 35000$ (1) $E_w = 5 \cdot 6 \times 10^{-15}$ J	2
		(ii)	Original $E_k = \frac{1}{2} mv^2$ (1) $E_k = \frac{1}{2} (1 \cdot 673 \times 10^{-27})(1 \cdot 2 \times 10^6)^2$ (1) $E_k = 1 \cdot 20 \times 10^{-15}$ (J) New $E_k = 1 \cdot 20 \times 10^{-15} + 5 \cdot 6 \times 10^{-15}$ (J) New $E_k = 6 \cdot 8 \times 10^{-15}$ (J) (1) $E_k = \frac{1}{2} mv^2$ $6 \cdot 8 \times 10^{-15} = \frac{1}{2} (1 \cdot 673 \times 10^{-27})v^2$ (1) $v = 2 \cdot 9 \times 10^6$ m s^{-1} (1)	5
	(b)		Alternating voltage has constant frequency (1) **OR** As speed of protons increases, they travel further in the same time. (1)	1
9.	(a)	(i)	$\Delta m = 4 \times 1 \cdot 673 \times 10^{-27} - 6 \cdot 646 \times 10^{-27}$ $\Delta m = 4 \cdot 6 \times 10^{-29}$ (kg) (1) $E = mc^2$ (1) $E = 4 \cdot 6 \times 10^{-29} \times (3 \cdot 00 \times 10^8)^2$ (1) $E = 4 \cdot 14 \times 10^{-12}$ J (1)	4
		(ii)	1 kg hydrogen has $\dfrac{0 \cdot 20}{1 \cdot 673 \times 10^{-27}} = 1 \cdot 195 \times 10^{26}$ atoms (1) Provides $\dfrac{1 \cdot 195 \times 10^{26}}{4} = 0 \cdot 2989 \times 10^{26}$ reactions (1) Releases $0 \cdot 2989 \times 10^{26} \times 4 \cdot 14 \times 10^{-12}$ $= 1 \cdot 2 \times 10^{14}$ J (1)	3
		(iii)	Large amount of energy released results in very high temperatures **OR** Strong magnetic fields are required for containment	1

Question			Answer		Max mark
	(b)		$m_{Rn}v_{Rn} = -m_a v_a$ (1) $3 \cdot 653 \times 10^{-25} \times v_{Rn} = 6 \cdot 645 \times$ $10^{-27} \times 1 \cdot 46 \times 10^7$ (1) $v_{Rn} = 2 \cdot 656 \times 10^5$ m s^{-1} (1)		3
10.	(a)		Blue light has higher frequency/ energy per photon than red light. (1) Photons of red light do not have enough energy to eject electrons (1)		2
	(b)		$E_k = hf - hf_0$ (1) $= (6 \cdot 63 \times 10^{-34} \times 7 \cdot 0 \times 10^{14}) -$ $2 \cdot 0 \times 10^{-19}$ (1) $= 2 \cdot 6 \times 10^{-19}$ J (1)		3
11.	(a)		Light with fixed/no phase difference.		1
	(b)	(i)	Bright fringes are produced by waves meeting in phase/crest to crest/trough to trough		1
		(ii)	$\Delta x = \dfrac{\lambda D}{d}$ (1) $\dfrac{9 \cdot 5 \times 10^{-3}}{4} = \dfrac{633 \times 10^{-9} \times 0 \cdot 750}{d}$ (1) $d = 2 \cdot 0 \times 10^{-4}$ m (1)		4
		(iii)	$\%uncert\Delta x = \dfrac{0 \cdot 2 \times 100}{9 \cdot 5 \times 10^{-3}} = 2 \cdot 1\%$ (1) $\dfrac{0 \cdot 002 \times 100}{0 \cdot 750} = 0 \cdot 27\%$ (1) Improve precision in (1) measurement of Δx		3
		(iv)	Green laser → shorter λ (1) Fringes closer together (1)		2
12.	(a)	(i)	Labels (quantities and units) and scale (1) Points correctly plotted (1) Correct best fit line (1)		3
		(ii)	Gradient of graph (1) Refractive index = 1·50 (1)		2
		(iii)	Repeated measurements Increased range of measurements Narrower beam of light Increase the number of values within the range Protractor with more precise scale eg ½° divisions		2
	(b)		$\sin \theta_c = \dfrac{1}{n}$ (1) $\theta_c = \sin^{-1} \dfrac{1}{1 \cdot 54}$ (1) $\theta_c = 40 \cdot 5°$ (1)		3
13.	(a)		$R = V/I$ (1) $= 12 / (30 \times 10^{-6})$ (1) $= 400\,000$ Ω (1)		3
	(b	(i)	$Q = It$ (1) $= 30 \times 10^{-6} \times 30$ (1) $= 900 \times 10^{-6}$ C (1)		3

Question			Answer		Max mark
		(ii)	$C = Q/V$ (1) $200 \times 10^{-6} = 900 \times 10^{-6} / V$ (1) $V = 4 \cdot 5$ V (1) Therefore voltage across resistor is $12 - 4 \cdot 5 = 7 \cdot 5$ V (1)		4
14.	(a)		Material 2		1
	(b)		resistance decreases (1) electron jumps (from valence band) to conduction band (1)		2

HIGHER PHYSICS 2015

SECTION 1

Question	Answer	Mark
1.	C	1
2.	B	1
3.	A	1
4.	D	1
5.	C	1
6.	B	1
7.	C	1
8.	E	1
9.	D	1
10.	B	1
11.	A	1
12.	D	1
13.	D	1
14.	D	1
15.	A	1
16.	E	1
17.	B	1
18.	D	1
19.	E	1
20.	C	1

SECTION 2

Question			Answer		Max mark
1.	(a)	(i)	**A** $v = 11 \cdot 6$ ms^{-1}	(1)	1
			B $v_h = 11 \cdot 6 \cos 40$		1
			$= 8 \cdot 9$ ms^{-1}	(1)	
			C $v_v = 11 \cdot 6 \sin 40$		1
			$= 7 \cdot 5$ ms^{-1}	(1)	
		(ii)	**A** $s = ut + \frac{1}{2}at^2$	(1)	4
			$4 \cdot 7 = 0 + \frac{1}{2} \times 9 \cdot 8 \times t^2$	(1)	
			$t = 0 \cdot 979$ (s)	(1)	
			Total Time $= 0 \cdot 98 + 0 \cdot 76$		
			$= 1 \cdot 7$ s	(1)	
			B $v = \frac{d}{t}$	(1)	3
			$8 \cdot 9 = \frac{d}{1 \cdot 7}$	(1)	
			$d = 15$ m	(1)	
	(b)		kinetic energy is less	(1)	2
			(as θ increases) speed decreases	(1)	

Question			Answer		Max mark
2.	(a)		(Total momentum before = total momentum after)		3
			$m_x u_x + m_y u_y = m_x v_x + m_y v_y$	(1)	
			$(0 \cdot 25 \times 1 \cdot 20) + (0 \cdot 45 \times -0 \cdot 60) = (0 \cdot 25 \times -0 \cdot 80) + (0 \cdot 45 \times v_y)$	(1)	
			$0 \cdot 30 - 0 \cdot 27 = -0 \cdot 20 + 0 \cdot 45 \times v_y$		
			$0 \cdot 45 \times v_y = 0 \cdot 23$		
			$v_y = 0 \cdot 51$ ms^{-1}	(1)	
			(to the right)		
	(b)	(i)	impulse = area under graph		3
			$\left(= \frac{1}{2} b \times h\right)$	(1)	
			$= \frac{1}{2} \times 0 \cdot 25 \times 4 \cdot 0$	(1)	
			$= 0 \cdot 50$ N s	(1)	
			Accept $0 \cdot 5$, $0 \cdot 500$, $0 \cdot 5000$		
		(ii)	$0 \cdot 50$ kg ms^{-1}	(1)	1
		(iii)			3
			Constant velocity at correct values and signs before <u>and</u> after collision	(1)	
			Velocity change from initial to final in $0 \cdot 25$ s.	(1)	
			Shape of change of velocity correct ie initially gradual, increasing steepness then levelling out to constant velocity.	(1)	
3.	(a)		$F = \frac{GMm}{r^2}$	(1)	3
			$F = \frac{6 \cdot 67 \times 10^{-11} \times 6 \cdot 42 \times 10^{23} \times 5 \cdot 60 \times 10^3}{\left(3 \cdot 39 \times 10^6 + 3 \cdot 70 \times 10^6\right)^2}$	(1)	
			$F = 4 \cdot 77 \times 10^3$ N	(1)	
	(b)		$g = \frac{W}{m}$	(1)	3
			$g = \frac{4770}{5600}$	(1)	
			$g = 0 \cdot 852$ N kg^{-1}	(1)	

Question			Answer		Max mark
4.	(a)		photons of particular/some/ certain energies/frequencies are absorbed (1)		2
			in its/the <u>Sun's</u> (upper/outer) atmosphere/outer layers (1)		
	(b)	(i)	light is redshifted/ shifted <u>towards</u> red (1)		2
			(as) the galaxies are moving away (from the Sun) (1)		
		(ii)	$z = \dfrac{\lambda_{observed} - \lambda_{rest}}{\lambda_{rest}}$ (1)		2
			$= \dfrac{450 \times 10^{-9} - 410 \times 10^{-9}}{410 \times 10^{-9}}$ (1)		
			$= 0 \cdot 098$		
		(iii)	$z = \dfrac{v}{c}$ (1)		5
			$0 \cdot 098 = \dfrac{v}{3 \cdot 00 \times 10^{8}}$ (1)		
			$(v = 2 \cdot 94 \times 10^{7}\ ms^{-1})$		
			$v = H_0 d$ (1)		
			$2 \cdot 94 \times 10^{7} = 2 \cdot 3 \times 10^{-18} \times d$ (1)		
			$d = 1 \cdot 3 \times 10^{25}\ m$ (1) $(1 \cdot 4 \times 10^{9}\ ly)$		
5.			Demonstrates no understanding 0 marks Demonstrates limited understanding 1 marks Demonstrates reasonable understanding 2 marks Demonstrates good understanding 3 marks		3
			This is an open-ended question.		
			1 mark: The student has demonstrated a limited understanding of the physics involved. The student has made some statement(s) which is/are relevant to the situation, showing that at least a little of the physics within the problem is understood.		
			2 marks: The student has demonstrated a reasonable understanding of the physics involved. The student makes some statement(s) which is/are relevant to the situation, showing that the problem is understood.		

Question			Answer		Max mark
			3 marks: The maximum available mark would be awarded to a student who has demonstrated a good understanding of the physics involved. The student shows a good comprehension of the physics of the situation and has provided a logically correct answer to the question posed. This type of response might include a statement of the principles involved, a relationship or an equation, and the application of these to respond to the problem. This does not mean the answer has to be what might be termed an "excellent" answer or a "complete" one.		
6.	(a)		Photon (1)		1
	(b)	(i)	$126\ GeV = 126 \times 10^{9} \times (1 \cdot 6 \times 10^{-19})$ (1) $= 2 \cdot 0 \times 10^{-8}\ (J)$ $E = mc^2$ (1) $2 \cdot 0 \times 10^{-8} = m \times (3 \times 10^{8})^2$ (1) $m = 2 \cdot 2 \times 10^{-25}\ (kg)$		3
		(ii)	$(2 \cdot 2 \times 10^{-25} / 1 \cdot 673 \times 10^{-27} =)\ 130$ (1)		2
			(Higgs boson is)		
			<u>2</u> orders of magnitude <u>bigger</u> (1)		
7.			Demonstrates no understanding 0 marks Demonstrates limited understanding 1 marks Demonstrates reasonable understanding 2 marks Demonstrates good understanding 3 marks		3
			This is an open-ended question.		
			1 mark: The student has demonstrated a limited understanding of the physics involved. The student has made some statement(s) which is/are relevant to the situation, showing that at least a little of the physics within the problem is understood.		
			2 marks: The student has demonstrated a reasonable understanding of the physics involved. The student makes some statement(s) which is/are relevant to the situation, showing that the problem is understood.		

Question			Answer	Max mark
			3 marks: The maximum available mark would be awarded to a student who has demonstrated a good understanding of the physics involved. The student shows a good comprehension of the physics of the situation and has provided a logically correct answer to the question posed. This type of response might include a statement of the principles involved, a relationship or an equation, and the application of these to respond to the problem. This does not mean the answer has to be what might be termed an "excellent" answer or a "complete" one.	
8.	(a)		The power per unit area (incident on a surface)	1
	(b)		$134 \times 0.2^2 = 5.4$ $60.5 \times 0.3^2 = 5.4$ $33.6 \times 0.4^2 = 5.4$ $21.8 \times 0.5^2 = 5.5$ (2) Statement of $I \times d^2 = $ constant (1)	3
	(c)		$I \times d^2 = 5.4$ (1) $I \times 0.60^2 = 5.4$ (1) $I = 15$ W m^{-2} (1)	3
	(d)		Smaller lamp (1) Will be more like a point source (1) Or Black cloth on bench (1) to reduce reflections (1)	2
	(e)		$A = 4\pi r^2 = 4\pi \times 2^2 = 50.265$ (1) $I = \dfrac{P}{A}$ (1) $I = 24/50.265$ (1) $I = 0.48$ W m^{-2} (1)	4
9.	(a)	(i)	• Different frequencies/colours have different refractive indices (1) OR • Different frequencies/colours are <u>refracted</u> through different angles (1)	1
		(ii)	$n = \dfrac{v_1}{v_2}$ (1) $1.54 = \dfrac{3.00 \times 10^8}{v_2}$ (1) $v_2 = 1.95 \times 10^8$ ms^{-1} (1)	3

Question			Answer	Max mark
	(b)	(i)	$v = f\lambda$ (1) $3.00 \times 10^8 = 4.57 \times 10^{14} \times \lambda$ (1) $\lambda = 656.5 \times 10^{-9}$ $m\lambda = d \sin\theta$ (1) $2 \times 656.5 \times 10^{-9} = d \times \sin 19.0$ (1) $d = 4.03 \times 10^{-6}$ m (1)	5
		(ii)	• different colours have different λ (1) • $m\lambda = d \sin\theta$ (1) • (m and d are the same) • θ is different for different λ (1) OR • different colours have different λ (1) • Path difference = $m\lambda$ (1) • (for the same m) • PD is different for different λ (1)	3
10.	(a)	(i)	12.8 J (of energy) <u>is gained by/ supplied to</u> 1 coulomb (of charge passing through the battery)	1
		(ii)	$E = V + Ir$ and $V = IR$ (1) $E = I(R + r)$ $12.8 = I(0.050 + 6.0 \times 10^{-3})$ (1) $I = 230$ A (1)	3
		(iii)	(Wire of large diameter) has a low resistance (1) OR to <u>prevent</u> overheating (1) OR to <u>prevent</u> wires melting (1)	1
	(b)	(i)	12.6 V (1)	1
		(ii)	(gradient = $-r$) gradient = $(12 - 12.5)/(60 - 10)$ (1) $= -0.01$ (1) internal resistance = 0.01 Ω (1)	3
		(iii) (A)	$I = \dfrac{V}{R}$ (1) $= \dfrac{(15 - 11.5)}{(0.09 + 0.45)}$ (1) $= 6.5$ A (1)	3
		(B)	The e.m.f. of the battery increases (1) Difference between the two e.m.f.s decreases (1)	2

HIGHER PHYSICS 2016

Question			Answer	Max mark
11.	(a)		$C = \dfrac{Q}{V}$ (1) $64 \times 10^{-6} = \dfrac{Q}{2 \cdot 50 + 10^3}$ (1) $Q = 0 \cdot 16 (C)$	2
	(b)		$E = \dfrac{1}{2}QV$ (1) $E = \dfrac{1}{2} \times 0 \cdot 16 \times 2 \cdot 50 \times 10^3$ (1) $E = 200$ J (1)	3
	(c)	(i)	$V = IR$ (1) $2 \cdot 50 \times 10^3 = 35 \cdot 0 \times R$ (1) $R = 71 \cdot 4 \ \Omega$ (1)	3
		(ii)	The voltage decreases (1)	1
		(iii)	Smaller initial current (1) Time to reach 0 A is longer (1)	2
12.	(a)		Suitable scales with labels on axes (quantity and units) (1) [Allow for axes starting at zero or broken axes or an appropriate value e.g. 30°] Correct plotting of points (1) Smooth U shaped curve through these points. (1)	3
	(b)		36° and 66°	1
	(c)		37°	1
	(d)		Correct substitution into equation using D_m from answer to (c) (1) Correct value for n (1·5 if using D_m equal to 37°) (1)	2
	(e)		Repeat measurements (1) More measurements around/close to a minimum or smaller 'steps' in angle (1)	2

SECTION 1

Question	Answer	Mark
1.	B	1
2.	A	1
3.	C	1
4.	C	1
5.	C	1
6.	B	1
7.	C	1
8.	A	1
9.	E	1
10.	D	1
11.	D	1
12.	A	1
13.	C	1
14.	E	1
15.	C	1
16.	B	1
17.	D	1
18.	D	1
19.	E	1
20.	B	1

SECTION 2

Question			Answer	Max mark
1.	(a)	(i)	$u_v = 9 \cdot 1 \sin 24°$ $u_v = 3 \cdot 7$ m s^{-1} (1)	1
		(ii)	$u_h = 9 \cdot 1 \cos 24°$ $u_h = 8 \cdot 3$ m s^{-1} (1)	1
	(b)		$v = u + at$ (1) $0 = 3 \cdot 7 + (-9 \cdot 8)t$ $t = 0 \cdot 378$ (s) (total) $t = 0 \cdot 378 \times 2$ (1) (total) $t = 0 \cdot 76$ s OR $v = u + at$ (1) $-3 \cdot 7 = 3 \cdot 7 + (-9 \cdot 8) \times t$ (1) (total) $t = 0 \cdot 76$ s	2
	(c)		$s = v_h \times t$ (1) $s = 8 \cdot 3 \times 0 \cdot 76$ (1) $s = 6 \cdot 3$ m (1)	3
	(d)		Smaller displacement (1) curve with decreasing gradient (1)	2

Question			Answer		Max mark
2.	(a)	(i)	$\bar{d} = \dfrac{1\cdot31+1\cdot40+1\cdot38+1\cdot41+1\cdot35}{5}$ $\bar{d} = 1\cdot37\text{m}$ (1)		1
		(ii)	$\Delta\bar{d} = \dfrac{1\cdot41-1\cdot31}{5}$ (1) $\Delta\bar{d} = 0\cdot02\text{ m}$ (1)		2
	(b)		$\%\Delta m = \dfrac{0\cdot01}{0\cdot20}\times100 = 5\%$ (1) $\%\Delta h = \dfrac{0\cdot005}{0\cdot40}\times100 = 1\cdot3\%$ (1) $\%\Delta\bar{d} = \dfrac{0\cdot02}{1\cdot37}\times100 = 1\cdot5\%$ (1) Mass (has largest percentage uncertainty). (1)		4
	(c)	(i)	$E_p = mgh$ (1) $E_p = 0\cdot20\times9\cdot8\times0\cdot40$ (1) $E_p = 0\cdot78\text{ J}$ (1)		3
		(ii)	$E_w = Fd$ (1) $0\cdot78 = F\times1\cdot37$ (1) $F = 0\cdot57\text{ N}$ (1)		3
		(iii)	Any one from: All E_p converted to E_k All E_p converted to E_w Air resistance is negligible Ramp is frictionless Bearings in the wheels are frictionless The carpet is horizontal No energy/heat loss on the ramp etc.		1
3.	(a)		Total momentum before (a collision) is equal to the total momentum after (a collision) in the absence of external forces (1)		1
	(b)		$m_1u_1 + m_2u_2 = (m_1 + m_2)v$ (1) $(0\cdot85\times0\cdot55) + (0\cdot25\times-0\cdot3)$ $= (0\cdot25 + 0\cdot85)v$ (1) $v = 0\cdot36\text{ m s}^{-1}$ (1)		3
	(c)		$E_k = \frac{1}{2}mv^2$ ANYWHERE (1) Before $E_k = \frac{1}{2}m_Xv_X^2 + \frac{1}{2}m_Yv_Y^2$ $= (\frac{1}{2}\times0\cdot85\times0\cdot55^2)$ $+ (\frac{1}{2}\times0\cdot25\times0\cdot3^2)$ $= 0\cdot14\text{ (J)}$ (1) After $E_k = \frac{1}{2}mv^2$ $= \frac{1}{2}\times1\cdot1\times0\cdot36^2 = 0\cdot071\text{ (J)}$ (1) Kinetic energy is lost. (Therefore inelastic.) (1)		4

Question			Answer		Max mark
4.	(a)		$(0\cdot83 + 1\cdot20) - 1\cdot80$ (1) $0\cdot23\text{ m s}^{-1}$ (1)		2
	(b)	(i)	$3\times10^8\text{ m s}^{-1}$ or c (1) Speed of light is the same for all observers/all (inertial) frames of reference or equivalent (1)		2
		(ii)	$l' = l\sqrt{1 - \left(\dfrac{v}{c}\right)^2}$ (1) $l = 71\sqrt{1 - 0\cdot8^2}$ (1) $l = 43\text{ m}$ (1)		3
		(iii)	Any one from: Correct – from the perspective of the stationary observer there will be time dilation Incorrect – from the perspective of the students they are in the same frame of reference as the clock Not possible to say/could be both correct and incorrect – frame of reference has not been defined		1
5.	(a)	(i)	$\Delta X = 0\cdot04\text{ (m)}$ $X = 0\cdot016\text{ (m s}^{-1})$ (1) $\Delta Y = 0\cdot06\text{ (m)}$ $Y = 0\cdot024\text{ (m s}^{-1})$ (1)		2
		(ii)	More distant galaxies are moving away at a greater velocity/have a greater recessional velocity Or equivalent		1
	(b)		$z = \dfrac{\lambda_{observed} - \lambda_{rest}}{\lambda_{rest}}$ (1) $Z = \dfrac{667\times10^{-9} - 656\times10^{-9}}{656\times10^{-9}}$ (1) $Z = 0\cdot0168$ (1)		3

Question			Answer	Max mark
6.			This is an open-ended question. Demonstrates no understanding 0 marks Demonstrates limited understanding 1 marks Demonstrates reasonable understanding 2 marks Demonstrates good understanding 3 marks **1 mark:** The student has demonstrated a limited understanding of the physics involved. The student has made some statement(s) which is/are relevant to the situation, showing that at least a little of the physics within the problem is understood. **2 marks:** The student has demonstrated a reasonable understanding of the physics involved. The student makes some statement(s) which is/are relevant to the situation, showing that the problem is understood. **3 marks:** The maximum available mark would be awarded to a student who has demonstrated a good understanding of the physics involved. The student shows a good comprehension of the physics of the situation and has provided a logically correct answer to the question posed. This type of response might include a statement of the principles involved, a relationship or an equation, and the application of these to respond to the problem. This does not mean the answer has to be what might be termed an "excellent" answer or a "complete" one.	3
7.	(a)		$W = QV$ (1) $= 1.6 \times 10^{-19} \times 2000$ (1) $= 3.2 \times 10^{-16}$ J (1)	3
	(b)		$Q = It$ (1) $= 0.008 \times 60$ (1) $= 0.48$ (C) (1) $number = \dfrac{0.48}{1.6 \times 10^{-19}}$ $= 3.0 \times 10^{18}$ (1)	4
	(c)		Straight lines with arrows pointing downwards	1

Question			Answer	Max mark
8.	(a)		Mass is converted into energy	1
	(b)		$m_{before} = 3.3436 \times 10^{-27} + 5.0083 \times 10^{-27}$ $= 8.3519 \times 10^{-27}$ (kg) $m_{after} = 6.6465 \times 10^{-27} + 1.6749 \times 10^{-27}$ $= 8.3214 \times 10^{-27}$ (kg) $\Delta m = 3.0500 \times 10^{-29}$(kg) (1) $E = mc^2$ (1) $= 3.0500 \times 10^{-29}$ $\times (3.00 \times 10^8)^2$ (1) $= 2.75 \times 10^{-12}$ J (1)	4
	(c)		Plasma would cool down if it came too close to the sides (and reaction would stop)	1
	(d)		Up the page	1
9.	(a)		The waves from the two sources have a constant phase relationship (and have the same frequency, wavelength, and velocity)	1
	(b)		Waves <u>meet</u> in phase **OR** Crest <u>meets</u> crest **OR** Trough <u>meets</u> trough **OR** Path difference = $m\lambda$ Can be shown by diagram e.g. Diagram must imply addition of two waves in phase	1
	(c)		Path difference = $m\lambda$ (1) $0.282 - 0.204 = 2 \times \lambda$ (1) $\lambda = 0.039$ m (1) (39 mm)	3
	(d)		The path difference stays the same **OR** The path difference is still 2λ (1) because the wavelength hasn't changed (1)	2
10.	(a)		$n = \sin i/\sin r$ (1) $= \sin36/\sin18$ (1) $= 1.9$ (1)	3
	(b)		$\sin\theta_c = 1/n$ (1) $= 1/1.9$ (1) $= 0.5263$ $\theta_c = 32°$ (1)	3
	(c)		Completed diagram, showing light emerging (approximately) parallel to the incident ray	1

Question			Answer	Max mark
11.			This is an open-ended question. Demonstrates no understanding $\quad$ 0 marks Demonstrates limited understanding $\quad$ 1 marks Demonstrates reasonable understanding $\quad$ 2 marks Demonstrates good understanding $\quad$ 3 marks **1 mark:** The student has demonstrated a limited understanding of the physics involved. The student has made some statement(s) which is/are relevant to the situation, showing that at least a little of the physics within the problem is understood. **2 marks:** The student has demonstrated a reasonable understanding of the physics involved. The student makes some statement(s) which is/are relevant to the situation, showing that the problem is understood. **3 marks:** The maximum available mark would be awarded to a student who has demonstrated a good understanding of the physics involved. The student shows a good comprehension of the physics of the situation and has provided a logically correct answer to the question posed. This type of response might include a statement of the principles involved, a relationship or an equation, and the application of these to respond to the problem. This does not mean the answer has to be what might be termed an "excellent" answer or a "complete" one.	3
12.	(a)	(i)	$V = IR$ $\qquad$ (1) $V = 1 \cdot 80 (4 \cdot 8 + 0 \cdot 1)$ $\qquad$ (1) $V = 8 \cdot 82$ (V) $\qquad$ (1) Voltmeter reading $(= 12 \cdot 8 - 8 \cdot 82) = 4 \cdot 0$ V $\qquad$ (1)	4
		(ii)	(Reading on voltmeter)/(voltage across lamp) decreases $\quad$ (1) (Total) resistance decreases/ current increases. $\quad$ (1) Lost volts increases/V_{tpd} decreases/ p.d. across 4·8 Ω increases/<u>share</u> of p.d. across parallel branch decreases $\quad$ (1)	3

Question			Answer	Max mark
	(b)	(i)	(Voltage applied causes) <u>electrons</u> to move towards <u>conduction band</u> of p-type/ away from n-type (towards the junction) $\qquad$ (1) Electrons move/drop from conduction band to valence band $\qquad$ (1) <u>Photon</u> emitted (when electron drops) $\qquad$ (1)	3
		(ii) (A)	$E = hf$ $3 \cdot 03 \times 10^{-19} = 6 \cdot 63 \times 10^{-34} \times f$ $\quad$ (1) $f = 4 \cdot 57 \times 10^{14}$ (Hz) $V = f\lambda$ (1) for $\qquad$ both equations $3 \times 10^{8} = 4 \cdot 57 \times 10^{14} \times \lambda$ $\quad$ (1) $\lambda = 6 \cdot 56 \times 10^{-7}$ m $\quad$ (1)	4
		(ii) (B)	Red (1)	1
13.	(a)	(i)	12 V	1
		(ii)	$E = \frac{1}{2} C V^2$ $\qquad$ (1) $E = \frac{1}{2} \times 150 \times 10^{-3} \times 12^2$ $\quad$ (1) $E = 11$ J $\qquad$ (1)	3
	(b)		$(R_T = 56 + 19 = 75 \ (\Omega))$ $I = \dfrac{V}{R}$ $\qquad$ (1) $I = \dfrac{12}{75}$ $\qquad$ (1) $I = 0 \cdot 16$ A $\qquad$ (1)	3
	(c)		(Lamp stays lit for a) shorter time $\qquad$ (1) (As smaller capacitance results in) less energy stored/less charge stored $\qquad$ (1)	2
14.	(a)		$f = \dfrac{1}{2L} \sqrt{\dfrac{T}{\mu}}$ $= \dfrac{1}{2 \times 0 \cdot 550} \sqrt{\dfrac{49 \cdot 0}{4 \cdot 00 \times 10^{-4}}}$ $\quad$ (1) $= 318$ Hz $\qquad$ (1)	2
	(b)	(i)	Suitable scales with labels on axes (quantity and units) $\quad$ (1) [Allow for axes starting at zero or broken axes or an appropriate value] Points plotted correctly $\qquad$ (1) Best-fit straight line $\qquad$ (1)	3
		(ii)	230 Hz	1

Acknowledgements

Permission has been sought from all relevant copyright holders and Hodder Gibson is grateful for the use of the following:

Image © Roman Chernikov/Shutterstock.com (SQP Section 2 page 12);
Image © Daseaford/Shutterstock.com (2015 Section 2 page 18);
Image © Dario Lo Pregti/Shutterstock.com (2015 Section 2 page 30);
Image © Sri Balaji Associates (2016 Section 2 page 26).